Senior Guide to iPhone

Table of Contents

Introduction

I live abroad and I have bought an iPhone for my parents. The problem is that my parents didn't know how to use an iPhone and I couldn't find any good guide on the internet. That's why I decided to write this guide! Now I can finally chat and video call with my parents and grandparents with no problems. And this is priceless.

I will give you all the important information so you can use the iPhone in the best of ways even if you are not good at technology.

Getting started with your new iPhone can be as straightforward as ensuring the iPhone is fully charged, tapping the power button, and following the on-screen directions to get everything set up.

If your new iPhone is your first iOS smartphone after switching from Android, we're here to help you make the switch. Even if you've been moving from iPhone to iPhone for years, our iPhone guide contains hints and recommendations for settings to change and features to test that you may have ignored. This guide will walk you through the process of getting started, including how to set up your device, navigate around, and master some of the operating system's most important features.

History of iPhone

Apple started off making computers in the late 1970s, mainly for its own local market. In 1977, the first computers, the Apple I and II were released. These early models didn't use miniaturized hard drives but instead relied on magnetic tapes to store data. From this point on, Apple would concentrate on bringing costs down and improving their products without compromising the user experience.

It took Apple until the early 90s to gain a considerable market share in the computer industry, with their Macintosh computers being very popular. This was due to both their increasing popularity as well as a sales slump among competitors like IBM and Dell. However, they experienced difficulties in producing at a cost that was affordable for consumers. This was because they were trying to produce something that had computers at its foundation but was still easy enough for those inexperienced to use.

Although the company made a profit, its stock value declined significantly after the price of Apple's stock collapsed in 1996. In addition to this, there were rumors among investors that their profits would suffer from competition from Microsoft's Windows operating system. This made the board of directors put pressure on Apple's CEO to release something revolutionary and this is when the idea for a combination PC and phone was born.

It took quite some time before the first usable prototype for the iPhone was released. In 2002 Steve Jobs announced that Apple had hired an outside company to create a device that would revolutionize cell phones. They then had about a year to figure out how it would work, although an early version of it showed up at Macworld in 2003. After a few iterations, the iPhone was finally released on June 29th of 2007.

CHAPTER 1:

Important Terminology

Control Center

Control Center is an iOS operating system feature that offers iOS devices easy access to critical device settings, simply by swiping up from the bottom of the display. It is your one-stop shop for instant access to dozens of controls for iPhone features, such as media playback, brightness, volume controls, mobile connectivity, screen mirroring, etc. This quick-access menu provides rapid access to some of your iPhone's most utilized, and or useful features and settings, without you having to launch the individual applications.

Airplane Mode

Airplane mode is a feature found on the iPhone and most mobile devices. When you enable this feature, all wireless signals from your smartphone are blocked. When enabled, you'll see an airplane icon in the status bar at the top of your iPhone. Airplane mode allows you to turn off cellular, Wi-Fi, and Bluetooth connectivity.

Cellular Data

Cellular data utilizes the same network infrastructure that is used for cellphone calls, which is made available by cellphone towers, in order to connect you to the internet. Unlike Wi-Fi, cellular data is always available, provided you're within the coverage area of your mobile service provider.

Wi-Fi

This is the control for turning on the Wi-Fi connectivity of your iPhone. A Wi-Fi network is essentially an internet connection that is distributed by a wireless router to several electronic devices such as computers, tablets, smartphones, etc. It enables these devices to interface with the internet through the help of a wireless router.

Bluetooth

Bluetooth is a technology that allows data to be transmitted between devices across short distances.

Bluetooth waves can only travel short distances, and their frequency changes rapidly.

Media Playback

This control lets you manage running media files. You can pause or play a running media file, or skip a song to the next one, all from this panel.

Portrait Orientation Lock

Portrait Orientation Lock enables you to prevent your iPhone's display from flipping from portrait to landscape mode when the smartphone is tilted beyond a certain angle.

Do Not Disturb

This function mutes your iPhone, allowing you to hold your calls and other notifications while you sit through a meeting, eat, sleep, or work quietly.

All calls, messages and notifications on your iPhone are received and preserved quietly when in do not disturb.

Brightness Slider

The brightness slider helps you to manually control the brightness of your iPhone screen. If you force-touch on the brightness slider, you can also turn on or off True Tone, which self-adjusts your display brightness, depending on the ambient lighting condition of the immediate surroundings. There is also a control for Night Shift just beside the True Tone control switch, which is a feature that lets you manage the amount of blue light your iPhone's display emits.

Volume Slider

The volume control or slider, on the control panel, helps you regulate the volume of your device without having to push the volume rocker on the left side of the iPhone display.

Your device makes no noise, and its display does not brighten up when notifications come in, but you can still glance at notifications by physically turning the display on.

Screen Mirroring

Screen mirroring is a wireless way of reproducing what appears on one device on another device's screen concurrently.

Flashlight

The flashlight feature on your iPhone can be controlled from the control center.

The camera flash on your iPhone functions as a flashlight as well, which is a useful tool for improving vision in low-light situations. It is powered by the flash mechanism built into your iPhone's primary camera unit on the back of your device, and it sits near the camera's lens.

Timer

The timer, which is part of the clock application, can be used to count down from a specific time to zero. After you've set the timer, you can use other applications, or even push the Sleep button, to put the iPhone in Sleep mode. In the background, the timer will keep counting down, and it will give off a sound when the countdown is complete.

Calculator

The calculator application on your iPhone is a simple four-function calculating software for adding, subtracting, multiplying, and dividing. It can also function as a scientific calculator that can perform trigonometric and logarithmic calculations.

Camera

Tapping on the camera icon takes you directly to the camera app.

Other controls can be added to the control center, depending on your preference.

CHAPTER 2:

What You Need to Know

SOS Mode

How to set up emergency SOS calls and your medical ID to constantly check your health.

There are so many programming highlights incorporated into iPhone, some are adequately simple to find, others are somewhat more covered up however they are extremely valuable, yet really great for security as well.

Whenever arrangement, emergency SOS on iPhone doesn't simply give you a speedy choice to call crisis administrations on the off chance that you're in a difficult situation, yet it will likewise naturally caution your crisis contacts to your area.

There are a few distinct choices for setting up and actuating emergency SOS on the iPhone. We've gone through them all beneath to assist you with ensuring your iPhone is prepared to help you however much as could reasonably be expected assuming that you at any point feel perilous or need to contact crisis administrations.

Press the side/home button rapidly multiple times.

One of the highlights of the emergency SOS settings on the iPhone is the capacity to rapidly press the side button or home button on your gadget multiple times in progression to call crisis administrations.

To set this up: Open Settings > Emergency SOS > Toggle available to work with Side Button.

At the point when you press the side button or home button multiple times, an uproarious alarm will sound and a countdown from three will start. However, you can turn the count sound off. To switch it off: Open Settings > Emergency SOS > Toggle Off Countdown Sound.

Press and hold the side/home button and volume up or down.

Squeezing and holding the side button or home button on your iPhone and either the volume up or volume down will raise one more screen on iPhone.

The choices on this screen are Slide to Power Off, slide for Medical ID, and slide for Emergency SOS.

Assuming you slide the Emergency SOS segment, this also will call crisis administrations, however, it would be somewhat harder to do in the event that you were in a tough situation contrasted with squeezing the side or home button multiple times so it merits flipping on that component as above. Flipping it on doesn't handicap the capacity to likewise utilize the slide choice referenced in this segment.

Step By Step Instructions to Set Up and Alter Crisis Contacts on iPhone

It's feasible to add crisis contacts in Apple's Health application. They will be communicated something specific and with your most recent area when you use Emergency SOS to call crisis administrations.

To Set Up Crisis Contacts: Open Settings > Emergency SOS > Edit Emergency Contacts in Health. This will then, at that point, send off the Health application where you follow stages 4–9 underneath.

Then again, you can:

- Open the Health application
- Tap on your profile in the upper right corner
- Tap on Medical ID
- Proceed "Alter" in the upper right corner
- Tap on "Add Emergency Contact"
- Pick the contact
- Pick the number assuming they have multiple
- Select their relationship to you
- Tap "Done" in the upper right corner

Assuming you open Settings > Emergency SOS, you'll see the rundown of crisis contacts that will be communicated something specific and your area.

Instructions to Set Up Your Medical ID on iPhone

It merits setting up your Medical ID on your iPhone simultaneously with setting up Emergency SOS and contacts.

It's feasible to set it to permit anybody to get to your Medical ID from your iPhone Lock Screen assuming you're in a mishap, for instance, and the specialists need to know whether you have any sensitivities or any prior ailments.

Open the Health application > Tap on your profile in the upper right corner > Tap on Medical ID > Tap on Edit in the upper right corner > Fill in any circumstances and make any notes > Toggle on Show When Locked > tap "Done."

Siri

Siri is Apple's voice-controlled virtual assistant available on your iPhone, which lets you carry out simple tasks hands-free by simply issuing voice commands.

The concept is that you give directives to Siri like you would a personal assistant if you had one, and the built-in virtual assistant will help you carry out the task.

You can ask Siri anything, from simple weather questions to more in-depth inquiries about everything, from news updates to the amount of lycopene in tomatoes. Siri can also help you carry out tasks like booking a hotel reservation or delivering a message via SMS or iMessage.

Siri can also activate and deactivate settings, search for stored information on your device, set an alarm and a reminder, make calls and text messages, and perform a plethora of other tasks.

CHAPTER 3:

How Your iPhone Works

Language Setup

After switching on your iPhone, the screen will display a series of "Welcome" texts, one after another, in different languages. At the bottom of the "Welcome" screen, you will see a straight black line which you should tap and slide upwards. This will bring up different language options, with the English language being the first on the list. Choose the appropriate language and hit "Continue" to move to the next setup screen.

Select Your Country or Region

The above text is what you should see right after settling on your language choice. On this screen, you are prompted to choose your region or country. Meanwhile, the first suggested country might be the United States, but if you happen to be in any other region or location, you can make your selection from the list of outlined countries by tapping on it.

Quick Start

The Quick Start setup screen offers you an opportunity to bypass this setup process by giving you a way to sync any data you have on a previously owned iPhone or iPad. But, if you don't have either of those or don't see any reason to perform this action, you can skip it by tapping on the Set Up Manually, highlighted at the bottom section of the screen.

Activation Screen

After skipping the Quick Start screen, your iPhone will display the following message: "It may take a few minutes to activate your iPhone." A rotating animation shows up under this screen, before leading to the Privacy and Policy setup screen.

Privacy and Policy

Your iPhone will display the Data and Privacy screen next, and at the bottom of the screen, you'll find two buttons: Continue and Set Up Later In Settings. Hit Continue if you wish to set up the privacy and policy features, but if you want to skip this step instead, tap on Set Up Later in Settings. If you go ahead and tap Continue, the next screen that pops up is Face ID.

Face ID

Face ID is a privacy feature that protects your device from unauthorized handling and use. It also restricts unapproved access to your data and files on your device. To set up Face ID, tap on the Continue button and follow the on-screen directions to register your face for the Face ID security feature. The scanning will begin, showing you tips on the proper placement of your face to get it scanned and recorded.

Connect to Wi-Fi

Connect your internet through a Wi-Fi network connection.

Get your Apple ID and Password: Create an Apple ID during your setup if you don't have one. Input your debit or credit card information to add Apple Pay to your setup.

You will need a backup of your previous device to transfer your data into your new device.

Apps and Data

Outlined on this screen are several data transfer options, including:

- Restore from iCloud Backup
- Restore from Mac or PC
- Transfer Directly from iPhone
- Move Data from Android
- Don’t Transfer Apps and Data

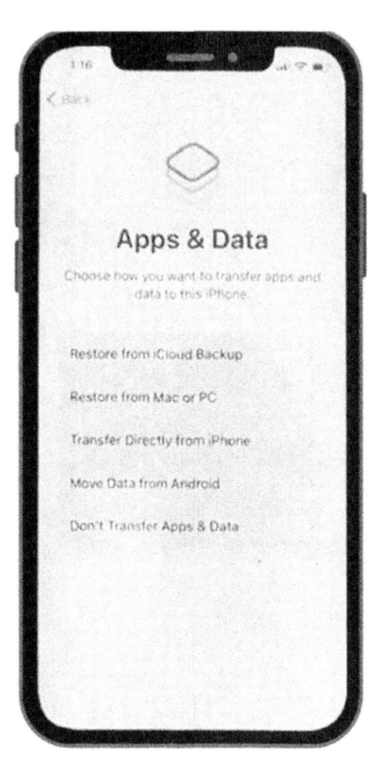

Depending on what action you take on the Apps and Data screen, by tapping on any option (except the Don’t Transfer Apps and Data option), you can transfer your old files and data from any listed device. Hence, you can use this mechanism to transfer your data from a cloud backup, or another device, to your new iPhone.

iCloud Backup Restore

The first option is to recover your data from the most recent iCloud backup, but you can also restore your prior applications and data from your PC. There's also an option to transfer all essential data from your previous iPhone right away. Another option is to move all necessary data from an Android smartphone to your new iPhone (if you had previously used an Android device).

You can, of course, also choose to skip the entire data transfer process.

Express Settings

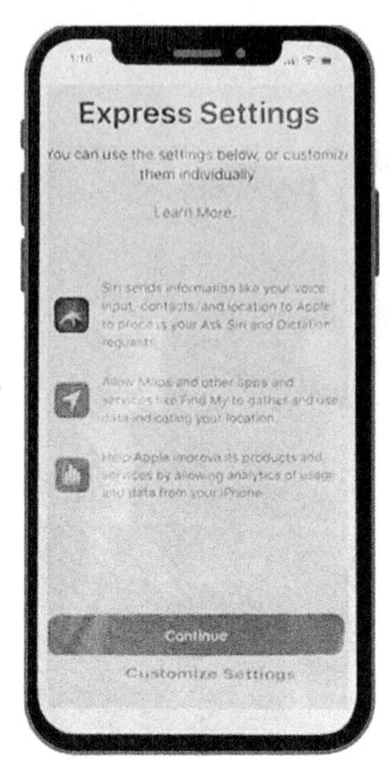

This screen displays when you've tapped on Don't Copy on the previous screen, with the assumption that your iPhone is brand new, and is not replacing a previous smartphone of yours.

This screen deals with setting up Siri, Maps, and granting specific device-related permissions to Apple (this is optional).

Tap on the Continue button at the bottom of this screen to proceed, or if you wish, you can customize these settings by tapping on Customize Settings.

Keep Your iPhone Up to Date

The update setup feature provides your iPhone with an auto-updating function, keeping your phone updated with new security features and operating system updates.

It can also send notifications to you about any available iOS updates.

Apple Pay

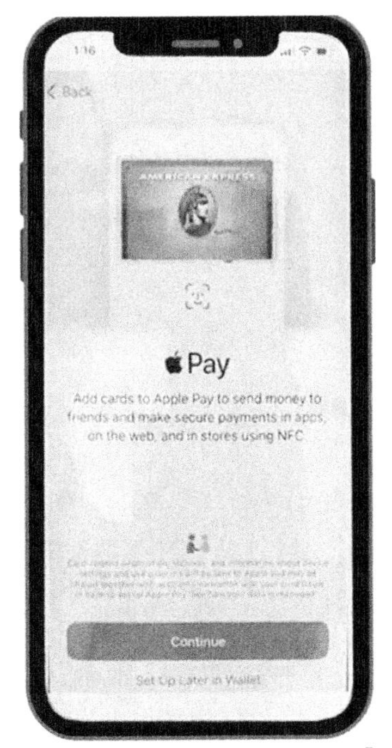

Apple Pay enables you to scan and store any credit or debit cards for virtual payments. All you need to do is place the card on a flat surface and then point your camera at the card. Make sure it's aligned inside the marked section as directed by your camera app, and your iPhone will scan the card and link it to Apple Pay for online payments. A list of different cards, such as American Express, MasterCard, Visa, and others, will all be displayed on the screen. If later, you need to pay for anything online, you won't be required to input any info about your card or to scan any cards.

The shortcut to accessing Apple Pay on your iPhone is to make a quick double-tap on your iPhone's power button.

Improve Siri and Dictation

The "Improve Siri and Dictation" setup screen is permission-required, asking if you want to share anything that has to do with your audio data with Apple.

Now, if you are comfortable with this feature, you can go ahead and tap on the "Share Audio Recording" option, but if not, you can tap on the "Not Now" option to skip this screen.

Using Screen Time

Follow the steps below to enable Screen Time:

Launch the Settings app.

Tap "Screen Time."

Select "Turn On Screen Time."

Tap Continue to proceed.

Select "This is My iPhone."

Appearance

The "Appearance" screen is the part of the setup process that provides you with the option of choosing either a light or dark screen display for your iPhone. You'll see the previews for both options. After selecting one, you can hit "Continue."

Display Zoom

The "Display Zoom" setup screen offers you the option to modify the sizes of icons or items displayed on your iPhone screen to your desire. You can make your selection of any display size that suits you best, and then proceed to the next screen. With all this done, you'll be met with the message: "Welcome to iPhone."

After all those initial steps in the setup process, your iPhone is now ready for use. After the previous step, a screen will be shown to you, displaying "Welcome to iPhone," notifying you that you are all done with setting up your device. Congratulations! You can then slide up to get started with more personalization. The next screen that should come up will be the "Home" screen.

Create a New Apple ID

Launch the Settings app.

Click on “Sign in to your iPhone” on the upper part of the screen.

Press “Don't have an Apple ID?”

Next, press “Create Apple ID.”

Now, input your birthday, then press "Next" at the top right of the screen.

Now input your name and press "Next."

Select your existing email address, or opt to “get a free iCloud address.”

Input your email address and then press "Next."

Following this, create a password that is eight characters long, and press "Next." Keep in mind that your password will have to include at least one uppercase letter and at least one numeral to be accepted.

You will receive a text message or call to confirm your identity, at which point you can press “Continue.”

Agree to the terms and conditions.

Enter your iPhone passcode, if it has one.

Select whether you want a confirmation email sent to the email address you entered or a different source.

Enter the verification code you received on your iPhone.

Next, press “Merge.”

That's all! From here, you can adjust the payment and shipping information, set up iTunes and the App Store, set up Family Sharing, and whatever else you're interested in.

Change Apple ID

To log in with another Apple ID on your iOS device, navigate to the "Settings" app.

Press on your name above. At the top of the screen, you can view the Apple ID you're logged in with. Scroll down and press "Sign Out."

Next, you will be required to enter your Apple ID and authorize your entry. In the following screen, you can select which information should be saved in the iCloud. To make a copy, you move the slider to the right; then, you tap again on "Sign Out."

Once your former Apple ID has been erased, you can now sign up again. Go into your settings and press "Sign in to iPhone." Now you can enter the email address and password for your other Apple ID.

Set Notification Preferences

You can choose whether to show an app notification on the lock screen or if you'd only like it shown when your face has been recognized.

Here's how to go about it:

Go to Settings.

Tap Notifications.

Now, press "Show Previews" to choose how content is or isn't shown on the lock screen. Otherwise, go to Settings, and then Notifications to adjust the lock screen look.

Allow Messages to Share Personalized Contact Data

You can create your very own contact image and name to appear on other people's iPhones. You can choose to turn on this option for just contacts, or for everyone; regardless, that person will have the last say on whether they acknowledge your chosen details.

Tap Settings.

Go to Messages.

Next, press “Share Name and Photo” where you can configure these and choose whom this automatically gets shared.

Enable Text, Call, and FaceTime Forwarding

For calls:

Launch the Settings app.

Tap Phone.

Next, tap “Calls on Other Devices,” and toggle on the switch for the devices you'd like to get calls on.

It's almost the same for messages:

Launch the Settings app.

Tap Messages.

Next, press “Text Message Forwarding,” which gets you to similar toggles for messaging.

CHAPTER 4:

Learn the Basics

Change the Keyboard Size

If you love typing with one hand, or you need to do it sometimes, there is an easy technique to make the keyboard smaller and move slightly to one side or the other.

If multiple keyboards are enabled, just tap and hold on the emoji or globe icon at the bottom left of the keyboard, and you will see the option to press the keyboard on the left or right side of your phone screen.

If you don't have many keyboards installed, you can enable this feature by going to Settings> General> Keyboard> One-Handed Keyboard.

To go back to the full size of the keyboard, just click the arrow to expand it again.

Make Bold

Did you know that you can create bold and italic text on the iPhone? Some apps even allow you to create monospaced, and strikethrough text.

While this doesn't work in all apps, you can change the text format in your email, notes app, and some third-party messaging and social media services (like WhatsApp).

To do this, click on the text, pick the passage you wish, and a copy-and-paste menu will appear, but it allows you to do much more than just copy-and-paste.

If the formatting functions are supported, you would see the "BUI" alternative for making changes in the pop-up that would show up.

Add a Contact

Tap the Contacts app to open it.

Tap ⊕ at the top right side of your phone.

Enter the name and other details of the contact.

Click ⊕ to enter the phone number of the contact.

Usually, Home will be the default option used for the contact's number. If you want to change this, select the arrow beside Home to explore other options (for example, mobile, work, and so on).

Click on Done to complete saving that contact.

Delete a Contact

Open the Contacts app.

Select the contact you would like to delete.

Select Edit on the top-right area of your screen.

Note that this does not stop the person from contacting you if that's what you were anticipating.

In order to do that, you would have to block them.

Update Existing Contact

You can make changes to an already existing contact. Here's how:

Open the **Contacts** app. Select the contact you would like to update. Select **Edit** on the top-right area of your screen.

Change the information that needs updating.

Click on **Done**.

Share a Contact

Follow the steps below to share your contact directly from the Contacts app:

Tap the contact you want to share on the **Contacts** app.

On the next screen, tap **Share Contact**.

On the Share sheet, choose the method you want to use to send the contact.

Send.

Block Contacts in the Phone App

Follow the steps below to prevent a contact from contacting you in the future:

Open the **Phone** app. Scroll down and tap either the **Recents** or **Contacts** button. If in the Recents tab, tap ⓘ by the side of the contact you want to block. If in the Contacts tab, click on the contact to open it.

Tap **Block This Contact**.

Tap **Block Contact** to complete your action.

Set Up App Limits

You can restrict the time you spend on specified apps, to give you time to do other important things. Once the set time is up, your phone will automatically block your access to the app for the rest of the day.

Launch the **Settings** app.

Tap **Screen Time**.

Select **App Limits**.

Then click on **Add Limit**.

Select all the app categories that you will like to place a limit on.

If you will rather limit certain apps within a category, click on the arrow by the right side of the category to display all the apps, then select the app that you want to limit.

Tap **Next**.

Then allot time for all the apps that you selected.

Select the **Days** that the rules should work.

Then click on **Add** to save your changes.

Control Center

This is a quick way to access the control center on your device:

Place your finger at the top-right edge of your screen, then swipe down to view the control center.

To close, tap on the screen, or swipe up from the end part of the screen.

Enable Control Center on Your Lock Screen

Would you like to be able to access the control center on a locked screen? Then follow the steps below:

Select **Face ID and Passcode** within the **Settings** app.

Type your phone's passcode when prompted.

Go down to **Control Center**, move the switch to the left to disable or to the right to enable.

Disable Access to Control Center from Within Apps

You can access the control center even when using apps on your phone. But if you do not want to be able to access the control center from within apps, follow the steps below to disable it:

Launch the **Settings** app.

Select Control Center.

Go to **Access Within Apps**, move the switch to the left to disable or to the right to enable.

Customize Control Center

By customizing your control center, you can add your frequently used apps and remove the apps you do not use often.

Select **Control Center** within the **Settings** app.

Select **Customize Controls**.

Tap ⊖ beside the apps you want to remove, then tap **Remove** to delete the controls.

Tap ⊕ beside the controls you want to add.

Rearrange Controls in the Control Center

Select **Control Center** within the Settings app.

Select Customize Controls.

Tap and hold the icon beside the controls you want to rearrange, then drag that to arrange them in the order that you like.

Modify Access to Items on Locked Screen

Select **Face ID and Passcode** within the **Settings** app.

On the next screen, go through the list and enable or disable the apps that you do not want to access on the locked screen.

Switch Between Open Applications

To quickly switch between open apps on iPhone with Face ID, slide the bottom edge of the screen left or right.

Multitasking Picture in Picture on iPhone

With Picture in Picture, you can use FaceTime or watch videos while using other applications.

A screen that shows the FaceTime conversation while you look at the Calendar app, fills the rest of the screen.

If you are using FaceTime or watching a video, tap Picture in Picture.

The video window shrinks to the corner of the screen so you can see the home screen and open other applications. In the video window that appears, you can do any of the following:

Resize the Video Window: Pinch to enlarge the small video window. To shrink again, squeeze it.

To Show or Hide Controls: Tap the video window.

To Move the Video Window: Drag to another corner of the screen.

To Hide the Video Window: Swipe from the left or right edge of the screen.

To Close the Video Window: Press Close.

To Return to the Entire FaceTime or Video Screen: Press the Full-Screen button in the small video window.

Type Using the Onscreen Keyboard on iPhone

In iPhone apps, you can type and edit the text using the onscreen keyboard. You can also use an external keyboard and dictation to enter text.

Change iPhone Sounds and Vibration

In the settings application, you can make changes to the sound your phone plays when you get an e-mail, text, reminder, call, voicemail, or other types of notifications.

Enter the Settings application > Sounds and Haptics.

Slide the slider under Ringers and Alerts to choose the sounds and volume.

To choose the vibration pattern & tones, touch a type of sound, like a ringtone or a text tone.

Do any of the below:

- Pick a tone (scroll to see all).
- Touch vibration, then select a vibration pattern or touch New Vibration to create one for yourself.

Change Your Wallpaper

On your device, select a picture or image as the wallpaper for your Lock and Home Screen. You can pick from live and still pictures. Enter the Settings application> Wallpaper> Choose New wallpaper. Do any of the below:

- Select a predefined image from the group at the upper part of your display (Dynamic, Stills, etc.).
- Choose any of your own pictures (touch an album, and touch the picture).
- Touch the Parallax Effect button, which would make the wallpaper move when you change your angle of view.

Touch Set, then choose either Set as Lock Screen, Home Screen, or Both

Apple Pay

It's simpler to utilize Apple Pay than utilizing a physical card and safer too. With your Wallet application cards, you can utilize Apple Pay for safe payments at Apple Pay shop, transit, applications, and compatible websites.

Adding a Card

In the wallet application, touch.

Pick any of the below:

- **Credit or Debit Card:** Arrange your phone in a way that your card is displayed in the frame on your screen or manually enter your card info.
- **Bus Cards:** Write your location or the name of your card, or scroll to check out the bus cards in your vicinity.

Your card issuer would determine if your card is qualified for Apple Pay and they might request more info to finish the verification process.

Set Your Default Card

The first card you set up in your Wallet would become your default card automatically. To choose a different card, adhere to the directives below:

In the wallet, pick your default card.

Long-press the card, then move it to the first card in the stack by dragging it there.

To change a card's position, hold down the card & move it to where you want by dragging it there.

Make Contactless Payments

With your debit, Apple Cash, and credit cards stored in the Wallet application, you can utilize Apple Pay for safe, contactless payments in stores, restaurants, etc.

You can utilize Apple Pay wherever you see one of these symbols:

To Make Payments with Your Default Card

Press the side button of your device two times.

When your card shows up, stare at your phone to confirm using your Face ID or write your login code.

Place the top of your iPhone within a few inches from the contactless reader till you see Done and the checkmark on your display.

Paying With a Different Card

When the default card appears, touch it and pick a different card.

Confirm using your login code or your Face ID.

Place the top of your iPhone within a few inches from the contactless reader till you see Done and the checkmark on your display.

Manually Adjust Your Screen Brightness

To change the brightness of your display, do any of the below:

Open the Controls Center and then slide the brightness button ☀.

Enter the Settings application> Display and Brightness, and slide the slider.

Automatically Adjust the Brightness of Your Screen

Your device can automatically change the brightness of your display to fit the current light conditions with an inbuilt ambient light sensor.

Enter the Settings application> Accessibility.

Touch Display and Text Size, and activate Auto-Brightness.

Create Folders on the Home Screen

Long-press the background of your home screen until the applications start jiggling.

You can create a folder by dragging an app onto another app.

You can add more applications by dragging them into the folder.

To change the name of the folder, long-press the folder, touch the Rename button and type the name you want.

If the applications start jiggling, touch the background of your Home screen and try again.

When you are done, touch the Done button, and double-tap the background of your Home screen.

You can delete a folder by dragging all the applications in it out.

Add a Widget to Your iPhone

Widgets display the current information from applications at a glance—today's headlines, battery levels, calendar events, weather, etc.

See Widgets Today

To view widgets in Today's View, swipe right from the left edge of your Home or Lock screen, and scroll.

Add a Widget to the Home Screen

Long-press the background of your Home screen until the applications start jiggling.

Touch the Add Widget button + at the upper left part of your display to open the Widget Gallery.

Scroll or search to look for the desired widget, touch it, and swipe through the size options.

Different sizes show different info.

When you find the size you like, touch the Add Widget button.

While the applications are still in jiggle mode, move the widget to where you want it on your display, and touch Done.

To remove a widget from your Home Screen, long-press it to display the Quick actions menu, and tap the **Remove Widget** button.

Move Applications and Widgets Around on Your Device

Long-press any application or widget on your Home Screen, and touch the Edit Home Screen button.

The applications would start jiggling.

Drag an application to any of the locations below:

- Anywhere on the page
- Another Home screen page

Drag the widget or application to the right edge of your display. You may have to wait a second for the new page to show. The dots on top of the dock indicate the number of pages there are and which of them you are in.

When you are done, touch Done.

Reset the Home Screen and applications to their original layout

Enter the Settings application> General> Transfer or Reset Phone.

Touch the Reset button, touch the Home Screen Layout button, and touch the Reset Home screen button.

All folders you created are deleted and the applications you download are sorted alphabetically after the apps that came with the device.

Uninstall Applications

Do any of the below:

Delete an Application from the Home Screen: Long-press the application on your Home Screen, touch Remove Application, and touch Remove from Home Screen to leave it in the Application Library, or touch Delete Application to delete it from your device.

Delete an Application from the Application Library and on the Home Screen: Long-press the application in the Application library, click the Delete Application button and touch Delete

Change or lock the screen orientation of your device

A lot of applications give you a different look when the iPhone is rotated.

You can lock the orientation of your display so that it does not change when the iPhone is rotated.

Open the Controls Center, then touch the Orientation Lock button .

Access more controls in the Controls Center.

A lot of controls have extra options. Long-press the control to view the options that are available. For instance, in the Controls Center you can do any of the below:

Long-press the control panel on the top left, and touch the AirDrop button to open the AirDrop options.

Long-press the camera key to record a video, take a selfie, etc.

You can personalize the Controls Center on your iPhone by adding more controls & shortcuts to many applications, like Notes, Voice Memos, Calculator, etc.

Enter the Settings application> Controls Center.

To add or remove a control, touch the add button or remove button beside a control.

To change the location of a control, click the Edit button beside a control and drag it to a new location.

Set Content and Privacy Restrictions

Go to Settings and tap Screen Time.

Press Continue and select "This is my [device]" or "This is my child's [device]."

If you are the parent or guardian of your device and want to prevent another family member from changing your settings, tap the Use screen time tag to create a password, then enter it again to confirm.

In iOS 13.4 or later, after confirming the password, you will be asked to enter your Apple ID and password. This will reset the screen time code if you forget it.

If you set the screen time on your child's device, follow the instructions until you get the parent code and the code. Re-enter the password to confirm.

In iOS 13.4 or later, after confirming the password, you will be asked to enter your Apple ID and password.

This will reset the screen time code if you forget it.

Airdrop

Send an Item Using AirDrop

AirDrop allows you wirelessly transfer your songs, videos, websites, locations, etc. to other devices and Mac computers nearby. AirDrop transmits information via WiFi and Bluetooth—both must be activated. To be able to use AirDrop, you must be signed in with your Apple ID. Transfers are encrypted for security reasons.

Open the item, then press the Share button

Tap Share .

Next, press AirDrop .

Here, click the **More** options button•••, or a different button that shows the app's sharing options.

Click on the AirDrop icon in the row of share options.

Proceed by pressing the profile image of a close AirDrop user.

If the person does not display as a nearby AirDrop user, request them to launch Control Center on their Apple device and permit AirDrop to receive items. To transfer to a Mac user, inform them to permit themselves to be discoverable in AirDrop in the Finder.

To send files via a method other than AirDrop, select the method from the row of sharing options. Siri can also recommend sharing methods by showing people's profile images and icons that signify sharing methods.

AirDrop can be used to securely share app and website passwords with someone using Apple devices.

Receive Files via AirDrop

Launch the Control Center, and press the AirDrop icon .

If you can't find the AirDrop icon, tap and hold the upper-left collection of controls.

Click on Contacts Only or Everyone to select the user you wish to receive the files from.

You can either approve or decline every single request.

Setting Up Google Mail, Calendars, and Contacts

Open the Settings app.

Next, tap **Mail**.

Tap **Accounts**.

Now, tap Add Account.

Choose **Google**.

Next, tap **Continue** if required to authorize Google.com to log in on your iPhone.

Fill in your Google account credentials.

On the interface, ensure the switches for contacts, mail, and other features are switched on or off.

Lastly, tap **Save**.

Setting up Outlook.com Mail, Calendar, and Contacts

Launch the Settings apps.

Tap **Mail**.

Tap **Accounts**.

On the interface, tap **Add Account**.

Next, press **Outlook.com**.

Input your Outlook.com account details.

Ensure the switches for contacts, mail, and others are switched on or off based on your preference.

Lastly, tap **Save**.

Setting up Exchange Mail, Calendar, and Contacts

Launch the Settings app.

Tap **Mail**.

Choose **Accounts**.

Choose Add Account.

Next, tap **Exchange**.

Fill in your Exchange email address.

Tap **Next**.

On the interface, tap **Configure Manually**.

Fill in your Exchange account credentials if you intend to set up your account manually.

Tap **Next**.

Ensure the toggles for contacts, mail, and other features are switched on or off based on your preference.

Lastly, press **Save**.

How to Dictate Text

If you are not a fan of typing, you can dictate your messages and have them written out as text. Just tap on the microphone icon next to the spacebar and start speaking. Click on Done once you are through with your speech. To include punctuation, just say the word. For instance, for the sentence "Hi Josh, how are you?" you would say, "Hi Josh comma how are you question mark." You can even say "new paragraph" to add a new line.

If you can't find the microphone icon on your keyboard, you will need to enable dictation.

Head to Settings, then select General, and then Keyboards. Activate "Enable Dictation."

How to Use Keyboard Shortcuts

For example, typing out your email address that's about 25 characters long can be shortened to just a few words. That's a lot less work. You can customize your keyboard shortcuts for almost anything. To create a keyboard shortcut:

Head to Settings, tap General, and then tap "Keyboard." Tap on "Text Replacement"

You will see Apple's shortcut example "On my way!" using "omw" as the shortcut

Tap the plus icon to customize your shortcut and then type in a Phrase and Shortcut.

Tap "Save."

You can cancel by selecting "Text Replacement" to return to the list of currently saved custom replacements.

How to Delete Unwanted Shortcuts

Swipe to the left on the shortcut. A short swipe will display the "Delete" button and a long swipe will automatically delete the shortcut.

How to Add or Change Keyboard

There are hundreds of foreign-language keyboards on your iPhone with which you can type.

How to Change your Keyboard

iPhone gives you access to more than 80 keyboards in foreign languages. Here's how to use them.

Head to Settings, then to General, and then tap Keyboard.

Tap on Keyboards and then select “Add New Keyboard” from the keyboard page.

Find and select your desired keyboard, then tap "Done."

Your new keyboard will appear on your keyboard list, along with your default and Emoji keyboard.

How to Change to a Third-Party Keyboard

There are lots of keyboards to download in the App Store which gives you several options to interact with your device. Here's how to use them.

Head to the app store and download any keyboard of your choice.

Launch the keyboard app and follow any instructions that pop up. In some cases, you need to go to settings, find the keyboard, and enable features.

Head to Settings, then to General, and then tap on Keyboard. Select “Add New Keyboard” from the keyboard page

Search for your new keyboard in the Third-Party Keyboards section and tap it.

Tap the new keyboard from the list and if needed, allow access.

How to Use a New Keyboard after You’ve Changed to It

Launch an app like “WhatsApp” that you can type. Tap or tap and hold the globe-shaped button at the bottom to use it. Works both ways

---

How to Change Your Default Keyboard

Go to Settings, then to General, and then tap on Keyboard.

Tap on Keyboards and then from the keyboard page, tap "Edit."

Rearrange the keyboard. You can arrange them by dragging three horizontal lines on the right side of your screen. Drag the preferred keyboard to the top of the list.

Tap "Done."

Set Emails to Download on Schedule

If you do not want to refresh your email manually, you can schedule the emails to download at a specified time. This is a balance between the two steps above; while you will not have to refresh your mail app manually, you will also not get an instant update. This method will still help to achieve the end goal, which is to save battery life.

From the Settings app, click on Passwords and Accounts.

Then click on Fetch New Data.

Navigate to the bottom and choose your options. The longer the time between checks, the longer your battery life is preserved.

Set Up the Screen to Auto-Lock Sooner

Auto-locking your screen helps to save your battery life. As long as your phone has something it is displaying, it will be taking out of the battery life. While I will advise that you select any of the options that suit you, do not choose Never, as that will drain your battery life.

From the settings app, click on Display and Brightness.

Then click on Auto-Lock.

Select any of the options from 30 seconds to 5 minutes.

CHAPTER 5:

AppStore and Apps

How to Download Apps and Games

Tap on the app or game you searched for—it could be free or you'll need to purchase it. If it is free tap on Get it or tap on the price if it is paid. Next, activate Touch ID by double-clicking the side button for Face ID or placing your finger on the Home button.

Apps installed from the App Store either appear on your Home screen or on a subsequent screen of apps. First, you need to search for cool apps to get them:

Head to the App Store and tap on the magnifying glass at the bottom of your screen, the search button).

Type in the app you want to search for and tap the search button.

How to Close Apps

Head to the app and hold the long line at the end of your screen with a finger. Slide the line upwards to close the app.

How to Close Multiple Apps

Open your home screen and slide the screen upwards from the bottom with your fingers. You will see all the apps you opened running in the background. Slide each app upwards to close them.

How to Find an App

To find an application, use the search tab at the lower part of your screen.

You can as well search through categories like games, books, applications, and more if you do not know what exactly you are looking for.

To ease some stress, you can make use of the Siri voice control to do your search. Just hold the Home button till Siri beeps.

How to Buy, Redeem, and Download an App

When you click on an application, you will be asked to either download it for free or make a payment. If the application is paid, know that you will be purchasing with your payment details on your Apple ID.

However, if the app displays what looks like a cloud, this means you have previously installed the app. Hence, you can install it again for free.

App Store Settings

You can set up your app store with different options by going to Settings and then to iTunes and App Store. This allows you to:

View and edit your account.

Change your Apple ID password.

Sign in with a new or different Apple ID.

Subscribe and turn on iTunes Match.

Turn on automatic downloads for books, music, TV, shows, movies, and more.

Control Offload Unused Apps

Remove unused apps on your device to free up storage space. In addition to automatically removing apps when the remaining storage capacity is low, you can also remove any apps manually.

By selecting "iTunes Store and App Store" from "Settings" on your iPhone and turning on "Offload Unused Apps," apps that are not in use are automatically used when the iPhone storage capacity is low and will be removed.

Restrict Offload Unused Apps

Go to the **Settings** app.

Next, press **Screen Time**.

Next, press to turn on Content and Privacy Restrictions.

Now, press iTunes and App Store Purchases.

Tap Deleting Apps.

Follow the arrow to the next screen and press **Don't Allow**.

With that control set, no one can delete apps from your device from now on unless you explicitly lift the restriction.

Move Home Screen Apps

First, press and hold any icon.

Then, a thumbprint will be displayed on the upper left of the icon, as shown in the image below, and it will move like a wave.

The icon can now be moved.

Then press the icon you want to move with your finger.

Move your finger to the place you want to move without releasing your finger.

By the way, you can't place app icons anywhere on your iPhone, like Android smartphones.

There may be cases where you can use unofficial app icons, but basically, app icons are arranged in order from the top.

Moving Apps to Another Page

Long-press the app.

A thumbprint appears on all icons.

Keep pressing the icon you want to move with your finger.

Move to the page you want to move to.

Switch to the page you want to move and place an icon.

Create a Folder on the Home Screen

First, press and hold the icon.

To delete, move, or create a folder, you must first press and hold the icon, the thumbprint is displayed, and the icon is wavy.

You can now edit the home screen icons.

First, tap the icons you want to organize into a folder.

Then, a folder is automatically created.

Move and place the icon in the folder as it is.

Reset Icon Layout on Home Screen

Launch the **Settings** app. Next, press **General**

Lastly, press Reset Home Screen Layout.

How to Switch Between Apps

On previous iPhones, you had to invoke the quick app switcher to swipe back and forth between apps. With the new iPhone, you can do it a lot faster. Use your finger to touch the gesture area at the bottom of the iPhone display. Swipe from left to right to return to the previous app or swipe from right to left to return to the next app.

Calendar App

You can add calendar events to your iPhone using the Calendar app, and you will be notified when the event's date approaches. To add an event, open the Calendar app and either tap the event day or the Plus sign located below the battery indicator in the upper right corner of your display. The Apple Calendar app can also sync with your Google Calendar and settings to ensure that all of your events are accessible across many platforms.

Safari

Safari, Apple's stock browser for all of their devices, is located right next to the Phone app on the iPhone dock. Safari has matured into a formidable browser over the years since its first release. It also includes strong privacy and security capabilities that track programs that attempt to access your data. You can disable these app trackers to prevent additional data theft from your account.

Numbers App

Numbers is a spreadsheet program created specifically for iOS-enabled mobile devices. The program features Multi-Touch gestures and the Smart Zoom function, allowing you to create excellent spreadsheets with just your fingers.

It is the second of three apps that comprise the iPhone's iWork productivity package.

Wallet App

The Wallet app keeps sensitive financial information, like credit card information, so you can conveniently make online transactions without having to constantly pull out your credit card.

Podcasts

It is Apple's podcast app, which comes preloaded on iOS devices and allows you to listen to your favorite podcasts from all around the world.

iMovie

On your iOS device, you may use the iMovie app, a video editing software tool, to create, edit, and export high-quality movies and trailers. The program can be used to change and improve the color configuration of videos, as well as their playback speed, and to stabilize shaky images.

YouTube App

Although the YouTube app does not come preloaded on your iPhone, I urge that you download it because it is a wonderful resource for content-rich, instructive, and enlightening films.

iTunes

The iTunes app, which is Apple's client application for the iTunes Store, is an online music store, media player, and music library where registered users can purchase music using a credit card. It also features Internet radio channels where users may listen to the greatest of Internet radio.

Stocks App

The Stocks app, as the name suggests, allows you to monitor stock values in real-time.

You may also look at the market capitalization of various stock exchanges throughout the world, as well as their gains and losses, transaction volume, and so on.

Camera App

The iPhone camera app has been updated to take advantage of Apple's latest features such as night mode portraits and Dolby Vision HDR recording. Below is the best way to take advantage of what the Camera app has to offer.

The cameras and the camera app are full of features, much of which Apple doesn't tell you about. Although there are several features and various ways to use them. Each feature is designed to make taking pictures more convenient, efficient, and faster.

Photos App

All of the images you've taken with your iPhone will be displayed in the Photos app.

You can also perform basic editing and erase obsolete or unattractive photos that you no longer require.

FaceTime

FaceTime is an Apple video and audio calling program that allows you to call your friends, family, acquaintances, and so on from anywhere on the globe for free. It uses network data and is only available on iOS-enabled mobile devices and Mac desktops.

Clips

The Clips app for the iPhone is an iOS app that allows you to make and share fascinating films with text, special effects, graphics, and so on.

Clock App

Alarms can also be established by tapping on the alarm icon or the Plus symbol just below the battery icon in the upper right corner of the alarm screen.

You can then specify the hour and minute that the alarm should sound, as well as whether or not it should be a repeating alert. You may program the alarm to make a certain sound when it goes off.

You can also select whether or not you wish to use the snooze feature. The Alarm app also includes a stopwatch feature that functions similarly to a traditional stopwatch.

TV App

There is also a TV app that allows you to quickly find all of your favorite movies and TV series with a few finger presses. When you're on the go, the TV app gives you access to all accessible programming from your active streaming subscriptions and cable TV. It's like watching TV when you're not at home.

Messages App

The Messages app is located to the right of Safari on the iPhone dock. The Messages app provides access to text messages (also known as SMS) and iMessage, Apple's instant messaging service available only on Apple-owned smartphones.

The **Messages** application aids in the organization of all your messages in a single location. You can compose and send messages to saved contacts or phone numbers. The Messages app works as iMessage if the destination device is an iPhone or an iPad, but for other non-Apple mobile devices, it functions as a conventional text message application.

You can then type in an exact URL (which is an acronym for Uniform Resource Locator, colloquially called a web address) or a search term, as it doubles as a search bar.

Now, if you want to save a Safari browser page to Favorites, or save it to your Safari browser bookmark, tap on the little arrow at the center of the icon row at the bottom of the browser screen.

This will bring up several options, and you can then choose whether to add the page to your **Favorites**, send the page to one of your contacts via iMessage, email the page, or add it to your Notes application. You can also bookmark the page, copy its contents, or add it to Safari's Reading List.

Music App

The music app is the last app on the dock of the iPhone. The iPhone Music app is a single-window software that lets you listen to all of your downloaded music as well as songs from your music library. It also has a search bar for locating the best music and playlists.

Pages App

It is an iPhone's word processing app that comes preloaded on the iPhone and allows you to create documents. You may collaborate with colleagues from anywhere on the globe using the Pages app.

Settings App

The Settings app manages everything on your iPhone, from how it works to what information it exchanges with Apple-owned and third-party apps. It also governs your security and external device connection procedures.

For maximum functionality, different applications necessitate distinct configurations and settings. The Settings app allows you to customize these distinct sets of settings to better serve the needs of each application.

Files

The Files app provides access to all of your files stored on online cloud services such as your iCloud account, Google Drive, Dropbox cloud storage, and any other cloud storage service, all in one place. You can browse the saved file directly on your iOS-enabled device and run a variety of commands on it.

The Health App

The Health app lets you track your fitness progress over time. You can also create a medical ID, which will contain a record of your medical condition and could be invaluable in an emergency.

Home App

The Home app connects to and securely controls HomeKit-compatible smart home devices. You can categorize your accessories by room, control several accessories at once, use Siri to control your smart home appliances, and so on from within the app.

Maps and Navigation

The Maps app comes next, which helps you identify locations and navigate more easily to places you've never visited before. It's excellent for holidays and sightseeing, but it might also save your life if you find yourself in a scenario where you need directions quickly in an emergency.

To use the Maps app, simply tap to activate it, enter a location in the search bar provided, and then tap Go. It will offer you step-by-step directions, much like a GPS.

Weather App

The Weather app displays the current weather conditions in your location. It also shows forecasted weather conditions for the next few days. You may also use the weather app to find out what the weather is like in specific regions across the world. The News app displays the most recent news stories from across the world. You can also use it to find location-specific news stories in any part of the world that pique your interest.

The Mail App

The Mail app is a preinstalled email service that lets you manage all of your email accounts from a single location on your device. With additional features to block unwanted email addresses and silence notifications from hyperactive discussions, you can take back control of your email account. You can compose professional emails while out and about using its high-level text formatting capabilities.

When you first launch the Mail app, you will be presented with a "**Welcome to Mail**" screen which has the icons of different email services.

Here, you can sign in to your email account. If, for instance, you use a Gmail account, you can tap on the Google icon when you launch the mail app to sign in to your existing Gmail account.

Once signed in, you can read emails, respond to received emails, and compose new emails all directly from the Mail app.

Phone App

On the bottom left of the iPhone dock is the **Phone** app. This is the app responsible for making and receiving voice calls. It has a **keypad** section that you can use to dial whatever number you are looking to call.

There is also a "Favorites" function in the **Phone** app (as shown below), which is a list of contacts that you frequently call and have saved to your Favorites for easy access when you want to call any of them.

Your new iPhone has no contacts saved in "Favorites" fresh out of the box. However, if you've imported your **Contacts** data, along with other important data, from a previous iPhone which had some contacts saved as Favorites, those contacts will appear in the "Favorites" section of your new device as well. If you want to add a new contact from your contact list to your Favorites, just tap on the blue + sign at the top left corner of the Favorites screen. The "**Recents**" section has a list of the recent calls that you've made, while the "**voicemail**" section has any voicemails your callers left on those calls.

The **Contacts** icon, which is located in the middle of the icon row at the bottom of the **Phone** app screen, contains a complete list of your phone number contacts. You can tap on any of the contacts to give them a call, view your call history with them, or add them to your favorites.

CHAPTER 6:

Everything About Your iPhone Camera

Open the Camera

Touch the Camera icon on your Home screen or swipe to the left on your Lock Screen.

Note: For protection, a green dot shows in the upper right part of your display when using the Camera.

Zoom

Pinch your display to zoom in or out. Or toggle between 0.5x, 1x, 2x, 2.5x, and 3x to quickly zoom. For more accurate zooming, long-press the zoom controls, and slide the slider to the right or left.

Snap a Macro Picture

You can take macro pictures and videos with the Ultra-Wide Camera, simply launch the Camera app, go close to the subject—as close as 2 cm—and your camera would stay in focus automatically.

Switch the Flash On or Off

Your device camera will use the flash if it needs to. To control the flash manually before taking a picture, simply touch the Flash button⚡.

Take a Picture with a Filter

Select Portrait mode or Photo mode, touch the Camera Cut button, and touch the Filter button.

Swipe right or left to view filters below the viewer; touch any of them to use it.

Use the Timer

Set the timer on the camera of your device to give yourself time to stay in position before the shot is taken.

Tap on the Camera Control button, touch the Timer button, select 3 or 10 seconds, and touch the Shutter button to start the timer.

Take a Live Picture

A live picture snaps what takes place before & after the shot, as well as the sound.

Open the Camera in Photos Mode

Ensure that you have activated Live Photo. When it's active, you will see the live picture button on the upper part of your camera.

Touch the Live picture button to enable or disable it.

Touch the Shutter button to take the picture.

To view the live picture, touch the photo's thumbnail at the lower part of your display, then long-press the display to play it.

Take Action Shots Using Burst Mode

You can utilize the Burst mode when capturing a moving subject. The burst mode captures many high-speed photos so you have many photos to pick from. You can take Burst photos with the back and front cameras.

Capture Rapid Pictures by Swiping the Shutter to the Left

The counter would show the number of shots taken.

Raise your fingers to stop

Touch the Burst thumbnail to pick the images you want to store and touch the Select button.

Touch the circle in the bottom-right part of any picture you would like to save as a separate picture, and then touch done.

Click on the thumbnail then touch the delete button to delete the whole Burst. Tip: You can long-press the volume up button to snap Burst pictures. Enter the settings application> Camera, and activate **Use Volume Up for Burst**.

Take a Picture With your iPhone Front Camera

Tap the change camera button .

Put your device in front of you, then touch the shutter

How to Take Portrait Photos

Head to the Camera app and select Portrait mode.

Follow the commands on your screen to frame the subject in the yellow portrait box.

Drag the Portrait Lighting control icon to choose a lighting effect:

Tap the "Shutter" to capture.

If you don't like an image you captured in Portrait mode, you can edit it by going to the Photos app, opening the photo, tapping on Edit, and then tapping on Portrait to turn the effect on or off.

How to adjust depth control in portrait mode

Go to "Camera," select Portrait mode, and then frame your subject.

Tap on "Depth Adjustment." The Depth Control slider will appear below the frame.

Drag the slider either to the left or right to adjust the effect. Tap on the "Shutter" to capture.

In Photos, you can further adjust the background blur effect of a Portrait image using the Depth Control slider.

Take Apple ProRAW Photos

The Apple ProRAW merges traditional RAW data with the image processing of your device to provide additional creative control when you customize the effect, color, and white balance. Apple ProRAW isn't compatible with Portrait mode.

To set up Apple ProRAW, enter the Settings application> Camera> Formats, and activate Apple ProRAW.

To take photos with Apple ProRAW:

Launch the camera application, then touch the ProRaw button RAW to activate ProRAW.

Take the shot.

When shooting, you can toggle between RAW and RAW.

Record a Video

Select video mode.

Press the record button, or press any of the volume buttons to start to record. You can do any of the below:

- Tap on the white shutter to snap a pic.
- Pinch your display to zoom.

Touch the record icon or press any of the volume buttons to end the recording.

Record a Slow-Motion Video

Videos recorded in Slo-mo mode, record as they should but the slo-mo effect can be seen when you play the video after recording. You can also adjust your video to start & stop the slow-motion effect at a certain time.

Select the Slo-mo re mode. Press the Record button or press any of the volume buttons to start recording. You can press the shutter to take a picture as you record.

Touch the record icon to stop the recording.

If you want to play a part of the video in slow motion & the rest at normal speed, touch the video thumbnail, and touch Edit. Drag the vertical line under the frame viewer to select the segment you want to play in slow motion.

Use the Live Text Feature with Your Camera

Your phone's camera can copy and share text, open sites, compose e-mails, and make phone calls from a text that appears on the camera.

Open the camera, and set your iPhone so that the text appears on the camera.

After the yellow frame shows around the visible text, press the text button.

Swipe or utilize the grab points to highlight the text, and do any of the below:

- Copy text.
- Select all.
- Look up: Search for text on the Internet.
- Translate.
- Share: Share the text via Messages, AirDrop, Mail, or any other available form.

To visit a site, call, or compose an e-mail, touch the site, phone number, or e-mail address on your display.

Touch the text key button to go back to the camera.

Scan QR Codes

You can utilize your camera to scan a Quick Response (QR) code for a link to a website, application, ticket, etc., your device camera would detect and highlight the QR code.

Open the camera, and set your Phone in a way that the code can be seen on the screen.

Touch the notification that shows on your display to go to the appropriate site or application.

Open the code scanner from the Controls Center.

Enter the Settings application> Controls Center, and touch the Add button beside the scanner code.

Open the Controls Center, touch the Code Scanner, and set your device in a way that the code shows on your display.

Touch the flashlight to switch it on.

Check Out Your Pictures

In the Camera, touch the thumbnail in the bottom-left part of your display.

Swipe to the right or left to see your latest photos.

Touch your display to hide or show the controls.

Touch All Photos to view all the pictures & videos saved in the Photo application.

Share Your Pictures

While viewing a picture, touch the Share button.

To share your photos, select an option, like AirDrop, mail, or messaging.

Take a Live Photo

Live Photo captures what happens just before and after the photo is taken, including sound.

Select Photo Mode.

Press the Live Photo button to turn Live Photos on or off.

Press the shutter button to take the picture.

You can choose to add effects to Live Photos, such as Loop and Bounce. See Edit live photos on the iPhone.

Take a Panoramic Picture

Use pan mode to capture landscapes or other images that do not fit on the camera screen.

Select Pan mode, then press the shutter button.

Rotate slowly in the direction of the arrow and hold it in the centerline.

Press the shutter button again to finish.

Press the arrow to pan in the opposite direction. To pan vertically, rotate the iPhone in landscape orientation. You can also reverse the direction of the vertical route.

Take Continuous Pictures

Burst mode takes multiple high-speed shots, so you can choose from a variety of shots. You can take continuous pictures with the rear and front cameras.

The counter shows how many shots you have taken.

Lift your finger to stop.

To select the images you want to keep, tap the Burst thumbnail, and then tap Select.

Below the thumbnails, gray dots indicate suggested images for preservation. Tap the circle at the bottom right of each photo you want to save as a photo, then tap Done.

To delete the entire series, tap the thumbnail, then tap Delete.

Enhance Images in Photos

Launch the **Photos** app.

Tap the photo you wish to enhance.

Tap **Edit** at the top of your screen.

Go to the bottom of the screen and tap the **Auto-Enhance** button .

Turn the dial to the right or to the left side to adjust the intensity of the image.

Tap **Done** to confirm your changes or tap **Cancel** to discard the changes.

Change Lighting in Your Photos

To manually change the lighting on your photos, follow the steps below:

Launch the **Photos** app.

Tap the picture you want to adjust.

Tap **Edit** at the top of the screen.

Go to the bottom of the screen and tap the **Auto-Enhance** button .

Swipe horizontally on the different menus and tap each category and adjust as needed.

When you make any adjustment, slide the icon to get a stronger or weaker effect.

When you finish, tap **Done** to complete.

Convert Photos to Black and White

Here is how to change your photo to Black and White:

Launch the **Photos** app.

Tap a photo to launch it.

Tap **Edit**.

Tap the button at the bottom of your screen.

Move left and right until you get to the three black and white filters: **Noir**, **Mono**, and **Silvertone**. When you get to each category, your image will automatically reflect the new color.

Slide the dial to modify the selected lighting aspect till you get a look that you like.

When you finish, tap **Done** to complete.

Start a Slideshow in Memories

You can convert your memories into a slideshow with the steps below:

From the **Photos** app, select the **For You** tab.

Scroll to the Memories section and select **See All**.

Click on the memory you want to access.

Tap the button on the cover of the memory to start the slideshow.

Change Your Slideshow Theme

You can change the theme of a slideshow with the steps below:

Follow the steps above to play the memory.

Then tap on any part of your screen while the memory is playing.

Tap the button to pause the slideshow.

Move down to the menu bar and swipe to the left or the right to view different themes. Click on the theme that you like. Each theme has its own different font face and background song.

When you find the theme that you like, tap it to apply it to your slideshow. Tap the button at the bottom of your screen to continue playing your slideshow.

Save a Memories Slideshow

After editing and modifying your slideshow, the steps below will show you how to save the slideshow:

Follow the steps above to edit your slideshow.

Once done, tap the icon at the bottom of your screen.

Look through the Share Sheet and tap **Save Video**.

Delete a Memory

Did Apple create a memory that you do not want to keep? The steps below will show you how to delete it:

From the **Photos** app, select the **For You** tab.

Scroll to the Memories section and select **See All**.

Tap the memory that you want to delete. Tap the icon at the top of your phone screen. **Now tap** Delete Memory.

Share a Memory

From the **Photos** app, select the **For You** tab.

Scroll to the Memories section and select **See All**.

Tap the memory that you want to view.

With the movie playing, click the icon and click on the method for sharing the video.

Share Photos or Videos

Follow the steps below to share an image or photo on your photo app.

From the **Photos** app, select the video or photo you want to share. Click the icon and click on the method for sharing the video or photo.

Share Multiple Videos or Photos

Here is a guide to sharing more than one image:

Launch the **Photos** app. Go to the top of your screen and tap **Select**.

Tap all the videos or photos you want to send out.

Click the icon and decide on the method for sharing the videos or photos.

Print Photos

From the **Photos** app, tap the image you want to print.

Click the icon at the bottom of your screen.

Tap **Print** from the available options.

Click **Select Printer** and set up your printer.

Tap the − or + icons to input the number of copies you want.

Tap **Print** to print the image.

Shoot Video with Your iPhone

Launch the **Camera** app.

You can either tap **Videos** at the bottom of your screen or swipe to the right to enter the Video mode.

Tap to begin shooting your video.

Once done, tap button to stop recording.

Your video will now save in the Photo Library.

Save Live Photos as a Video

From the **Photos** app, tap the Live Photo you want to convert to Video. You will see the LIVE tag beside every Live Photo.

Click the icon and tap **Save as Video**.

The video will automatically save in the **Recents** album.

Create a Time-Lapse Video

Shoot this video to show the amount of time that passed since you started recording. To get the best result, let your phone be on landscape mode and also use a tripod.

Launch the **Camera** app.

Swipe to the right of your screen three times. Another way to get to the Time Lapse view is to click on **Time Lapse** after you make the first swipe.

Tap to begin shooting your video.

Once done, tap button to stop recording.

You get the best results when you shoot a long video.

Take a Still Photo While Capturing Video

Do you know that you can take a picture while shooting a video? Follow the steps below:

From the home screen of the **Camera** app, swipe to the right to get into the Video mode.

Tap to begin shooting your video.

While the video recording is on, tap the white button beside the Red button to capture your photo.

Change Wallpaper from the Photos App

Launch the **Photos** app and tap the image you want for your wallpaper.

Click the icon and tap **Use as Wallpaper**.

You will see a prompt on your screen to choose between **Live**, **Perspective**, **or Still Image**.

Tap **Set**.

You will receive a pop-up menu on your screen. Select the option you prefer.

Change Your Wallpaper

This is how to change your wallpaper:

Click on **Wallpaper** in the **Settings** app.

Select Choose a New Wallpaper.

Select the type of wallpaper you want from the options on your screen: **Photo Library**, **Still**, or **Dynamic**.

Tap **Photo Library** to select an image from your photo library.

Click on **Still** if you want a non-changing photo from Apple's photo library.

Click on **Dynamic Wallpaper** to get your images (that have effects) from Apple's photo library.

Tap a photo to go to Preview mode.

Move the photo around your screen until you are able to get the desired fit.

Tap **Set**.

You will receive a pop-up menu on your screen. Select the option you prefer.

Burst Shot

How to Activate Burst Mode

- Launch the **Camera app** and press the **Shutter** button.
- Set the **Burst Mode switch by sliding** your phone to left (portrait) or upwards (landscape).
- When you're done, tap the OK button immediately after releasing the **Shutter**.
- To return to normal mode, tap the Camera button again.

Pano Pictures

To start, you must select the Camera app from the app drawer and swipe right to open the camera view.

At the bottom, you will find a camera control button.

When you press that button, you will get to the camera settings page where you can set up panorama mode

Doing a QuickTake Video

The process is easy. You simply long-press the capture button on the app and your iPhone will take a picture and record it as a video. You can then share the video directly with social media apps, or email the video directly to a friend or family member.

How to Use the Front-Facing Camera

Selfies are a thing these days. It is awkward using your main camera instead of your front-facing camera for a selfie. To use your front-facing camera, open your phone camera and click on the camera icon with two circular arrows around it at the bottom of your screen. You will observe the change in camera modes as your face will show on your screen as it appears in the camera.

How to Adjust the Exposure

The exposure of the picture determines how much light the subject of the picture will receive. It helps emphasize certain aspects of the picture. To adjust the exposure of the picture, click on the subject on the picture and swipe up or down to adjust the exposure.

Swipe up to brighten the picture and swipe down to darken the picture.

Adjust the Camera Focus and Exposure

Before a photograph is taken, the iPhone camera automatically sets the concentration and exposure, and face identification adjusts the exposure of numerous faces. Do the following in the event you intend to manually adjust the focus and exposure:

Tap the screen to display the autofocus area and exposure settings.

Tap where you want to move the focus area.

Next to the focus area, pull the Adjust exposure button either up or down to adjust the exposure.

Touch the Camera Control button, touch the Exposure button, and then move the slider to adjust the exposure. The exposure will be locked until the next time you open the camera. To keep the exposure control so that it does not reset when you open the Camera, go to Settings> Camera> Keep settings and turn on Exposure adjustment.

CHAPTER 7:

Entertainment with Music, Videos, and News

One of the most popular iPhone functions is music. The device comes with a wide range of apps that can be used to listen to music on the go.

It also makes it easy to find and play your favorite songs and albums, or discover new ones. There are hundreds of thousands of free downloads available in iPhone's Music and Videos that are worth exploring.

If you are an average senior citizen with a low budget, then the iPhone is a great way to enjoy your retirement. Most apps are free or inexpensive to download and the phone itself is not too expensive. This user guide will help you understand how to use the iPhone for music videos and the latest news.

Install Apple TV App on the iPhone

With the Apple TV app, you can watch original programs and movies, plus your favorite or favorite programs, movies, games, and animated content with Apple TV+. You can include watch articles from Apple TV stations like Paramount + and showtimes, streaming services and cable providers, and buy or rent movies and TV shows.

The Apple TV app is on the iPhone and other Apple products and streaming devices, so you can watch it at home or anywhere.

Note: Apple TV app availability and its features and services (such as Apple TV+, Apple TV stations, games, news, and supported applications) vary by country or region.

Subscribe to Apple TV Stations

If you subscribe to Apple TV channels (such as Paramount + and Showtime), you can stream non-commercial content when you want or download it while you're on the go. If you use Family Sharing, about five family or family members can share the subscription with no additional charge.

Click View Now, browse the list of channels, then tap the channel you want to view.

Tap on the Subscription button, check out the free trial (if applicable) and registration details and follow the onscreen instructions.

Install a Cable or Satellite Service on an Apple TV Device

Individual signatures provide instant access to the supported non-supported software in your subscription package.

Go to Tools & TV > TV Providers.

Select your TV provider, then sign in with the details of your provider

If your TV provider is not listed, log in directly from the device you wish to use.

Devices with Apple News

Apple TV promotes the latest news or episode in the App Series that you watch on a connected device (supported apps only).

Click View Now, Streaming Tools, and Read Available Apps

To connect a device, tap it, and follow the onscreen instructions.

Manage your connected apps, payments, and watch history

Click View Now, then tap the My Account button or your top-level image on the top right. Click on any of the following:

Connect new or disconnect supported streaming devices.

Devices N devices appear on Apple TV devices on your devices where you log in with your Apple ID.

Manage Investment: Tap to change or cancel your subscription.

Clear Play History: Clear your visual history from your Apple and Apple devices.

Get About Programs and Movies

In the Apple TV menu, click Watch Now, and do one of the following:

- **Check Out the Following:** Line Up, Buy the Willow Title you just added, rent or buy, keep the next piece you are looking at in the series, or start what you are looking for after finishing.
- **Get Tips:** Find Will Watch The Watch Line author for the author's recommendations that you have made for everyone. This device offers many rules and regulations unique to your device based on your channel subscriptions, supported apps, pricing, and viewing interests.

- **Read Apple TV+:** In the Apple TV+ line, tap on the title to see more information or to play the trailer.
- **Visit Station:** Find this Sc to see the stations you subscribe to. In the row of channels, browse for the available channels, then tap a channel to browse for its topics.
- **Watch Live News:** Click on Top TV Programs, find the news bar down, and select a news channel (available in select countries or territories).
- **View Movies, Programs, and Events Sent by Friends to Messages:** Find this by rolling the shared YouTube line. View Messages Use to receive and share messages with friends on the iPhone.

Check Games

You can read about sports or leagues, watch games, and get information about nearby games. Click Watch Now, then tap Games Up. Do one of the following:

- **Browse Games:** This sc was created to watch many sporting events, such as football, baseball, and basketball. To reduce your reading, find this roll and select a play.
- **Watch the Game:** Tap the play.
- **Select your Favorite Groups:** Find this roll, and then click your favorite group.

Their games will be featured in the Up Next section, and you will receive notifications of your favorite bands. To hide most live games, go to Tools> TV, and turn off Show-Score scores.

Find Value Programs, Movies, and Games

Tap, villa, then enter your villa question. You can read titles, games, groups, cast members, Apple TV channels, or titles (such as "Car Chase").

Use the next unit list:

- **Enter One Item in the Next Unit:** Click on it to see its details, then click Add.
- **Remove Item Next Unit:** Hold item then tap Remove top.
- **Continue to Watch Another Device:** You can view your watch list next to your Apple iPhone, iPod, iPod Touch, Mac, Apple TV, or your supported TV where you are logged in.

Watch Programs and Movies on the iPhone on Apple TV

You can play programs and movies from the Apple TV device. Buying, renting, Apple TV+ and Apple TV stations run on the Apple TV listings, while news about proven providers goes on their video program.

Buy, rent, stream, or download programs and movies.

Tap on a thing to see its details.

Select one of these options (not all options are available for all topics):

- **Watch Apple TV+ or Apple TV Stations:** Tap Play. If you are not currently logged in, click try for free (available for eligible Apple ID accounts) or subscribe.

- **Select a Different Video Device:** If the title is available from many apps, apply this roll How to View and Select the device.

- **Buy or Rent:** Confirm your choice, and complete the payment.

Once you start watching a movie, you can play it as much as you want for 48 hours, after which the rental period will expire. When the rental expires, the movie will be canceled.

- **Download:** Click the Download button. You can read the library in your library and download it, even if it is not connected to the iPhone Internet.

- **Preview:** Check notes, then tap Before Race.

When something purchased in advance is available, your payment method is charged, and you will receive a notification via email. If you open Downloads automatically, things will pop up automatically on your iPhone.

Manage Playback on Apple TV Listings

During playback, tap the screen to show controls.

Change the settings of the Apple TV app.

Go to Tools> TV.

Select streaming option:

Use cellular data.

- **Cell:** Choose high quality or automatic.

- **Wi-Fi:** Choose high quality or data storage.

Requires a high-quality Internet connection and uses a lot of data.

Select Download Options.

Use cellular data.

- **Cell:** Choose high quality or download quickly.
- **Wi-Fi:** Choose high quality or download quickly.

High quality results in slow downloads and uses a lot of data.

Choose a Valley. Each additional audio increases the size of the download. To delete one of the delete wallets, roll the left-hand side of the merge to the one you want to delete, then tap Delete.

The origin is the main walleye of your country or region. If you have ordered AudioNote Tools> Accessibility, audio captions are also downloaded.

Based on your suggestions and the list of items listed above on the iPhone you're viewing on the iPhone, turn on the usage history.

What you see on the iPhone affects your unique recommendations on your iPhone and your next device on your device where you are logged in with your Apple ID.

Finish download.

Click Library.

Scroll to the left side of the item you want to delete, then tap Delete.

Deleting anything on the iPhone will not remove it that you purchased on iCloud. You can download the item on the iPhone later.

Manage Connected Apps and Subscriptions

Press **Watch Now**, and then choose the **My Account** button or your profile image from the upper right.

Click on any of the resulting options:

Connected Apps: Toggle applications on or off.

Connected apps display in the Apple TV app on each of your devices that you've logged in using your Apple ID.

Manage Subscriptions: Click on a subscription to switch or stop it.

Clear Play History: Erase your watching history from all your Apple gadgets.

Difference Between Automatic Setup or iTunes and Device-to-Device

When using Automatic Setup or iTunes, if you want to transfer all of your data to your new device, you would first need to make an iTunes backup or iCloud backup. And if you're on the iTunes road, you'll want to make sure it's an encrypted backup to successfully move all of your files, including confidential health details. For iCloud, if you just use the free 5 GB iCloud plan, you'll also want to make sure you have space for a backup, a difficult job.

Direct transfer from device to device is much less of a hassle.

If your iCloud storage space is almost complete, a direct transfer method is a great choice, and you need not back up your old iPhone before the migration. Your photographs, media, app data, settings, and more are included in the data that you transfer. To get the job done, the process utilizes a mix of Bluetooth and Wi-Fi.

Importantly, when app information is moved to your new smartphone, all your applications will be downloaded from the App Store again by your new iPhone, rather than only transferring them over from your older device. This is just how it works to recover from an iTunes backup, so this should be pretty familiar if you have experience with it.

How to Subscribe to Apple Arcade

Apple Arcade is a gaming subscription service on your iPhone device. It is also available on other Apple devices like the iPad, Apple TV, etc. At a monthly fee of $4.99 or a yearly fee of $49.99, you can download games from several available titles. You can also share all the downloaded games with members of your family sharing group.

Head to the app store and look down to the bottom you will see the arcade button.

You can do a free trial for a month. Click try free and agree to the "T&C."

Tap on Subscribe to start a monthly subscription. Review the subscription detail and confirm with your ID.

How to Cancel Your Apple Arcade Subscription

Go to App Store and tap on your profile icon.

Tap on Subscriptions, then on "Apple Arcade" and then tap on Cancel Subscription.

You can't play any Arcade games after you have canceled your subscription. You can re-subscribe to play the games and regain access to your gameplay data. You might lose some of your gameplay data if you don't re-subscribe on time.

Sign Up for Apple Arcade

Launch the **App Store**.

At the bottom of your screen, tap **Arcade**.

You have the first month free. Select **Try It Free** to begin the free trial.

Tap **Confirm** to begin your subscription.

Cancel Apple Arcade

Follow the steps below to unsubscribe from Apple Arcade:

Launch the **App Store**.

Select your Apple ID at the top.

Tap **Subscriptions**.

Tap Apple Arcade.

To stop the free trial, tap **Cancel Free Trial**, or tap **Cancel Subscription** to cancel the charged subscription.

Confirm your action.

How to Play Music

Click on the Music icon. This will open up the Library view. When you open the Music app for the first time, you may see a screen telling you to sign up for Apple Music. You can ignore and dismiss this for now.

Enter the Library interface.

Choose from any of these options: Playlists, artists, albums, songs, genres, compilations, and downloaded music.

You will also see Recently Added. Tap on songs, here you will see all the tracks.

How to Subscribe to Apple Music

Go to iTunes or the Apple Music app. Or go to music.apple.com to subscribe.

Go to Listen Now or For You and tap the trial offer.

Choose a subscription (individual, family, or student). You can share your family subscription with six people.

Sign in with your Apple ID or create a new one if you don't have one.

Confirm your billing details and add a payment method. Tap or click Join.

Tap iTunes and App Store Purchases

Select an option and set it to Disable.

You can also change the password settings for further purchases from iTunes and the App Store or Book Store. Follow steps 1–3. Step, and then select Always Request or Do Not Require.

Enable built-in applications and features

You can restrict the use of embedded applications or services. If you turn off an app or feature, it will not be deleted, it will only temporarily hide it from the Home screen. For example, if you turn off Mail, Mail only appears on the Home screen when you turn it on again.

Change permitted applications:

Select Settings> Screen time.

Touch Content and Privacy Restrictions.

Enter the screen time password.

Tap Allowed Apps.

Select the programs you want to activate.

Prevent explicit content and content rating

You can also prevent music with explicit content from being played, as well as movies or TV series with a specific rating. Applications also have configurable rankings with content restrictions.

CHAPTER 8:

Browser and Communication

Safari

Safari is the default web browser for all Apple devices. You can use it to view pages that are open on your other devices, share links, browse the web, and lots more. We will delve more into it in this chapter.

View and Reopen Recently Closed Tabs in Safari

Open the **Safari** app.

Click on the button at the bottom right side of your screen.

Tap and hold the new tab button until you see a list of the Recently Closed Tabs.

Click on a site to open the address in a new tab.

Tap **Done** to exit.

Customize Your Favorite Sites in Safari

On the Safari home page, you will find recommended websites, your favorite websites, frequently visited sites, and Siri suggestions. This guide will show you how to customize your favorite websites.

On the homepage of the Safari browser, under the **Favorites** section, click and hold a website's favicon to display the preview screen as well as the contextual menu. There are a couple of other options including **Edit** and **Delete**.

Tap **Edit** to rename the site as you want it to show on your Favorites.

In the website address field, you can also enter a different website to take you to a different part of that site.

Bookmark Multiple Open Tabs in Safari

Follow the steps below to bookmark different websites at once:

Open all the sites you plan to bookmark.

Let one of the websites be in the main browsing window.

Press long on at the bottom of your screen.

Click on **Add Bookmarks for X Tabs** on the next screen.

On the next screen, save the tabs in a new bookmark folder or choose from the current list and click **Save** at the top of the page to save your bookmarks.

Close All Your Open Tabs at Once

Follow the steps below to close all your open tabs at the same time:

Method 1:

Open the **Safari** browser.

Press long on at the right side of the bookmark icon.

Select Close All Tabs.

Method 2:

Tap once on icon to display the Window view.

Press long on **Done**.

Select Close All Tabs.

Automatically Close Safari Tabs

Set up your browser to close open tabs at a defined time.

Tap **Safari** in the **Settings** app.

Select Close Tabs.

Select your preferred option on the next screen.

Safari Share Sheet

Follow the steps below to share a web page as a link, archive, or PDF file:

Open the website you want to share.

Tap to display the Share Sheet.

Click your sharing method from the list. Tap **More** to see other options.

Select your sharing method and tap **Options** to choose to send as an archive, link, or PDF.

Enable Content Blockers in Safari

Content blockers offer a one-trick solution for prohibiting ads like popups and banners from stacking on websites you visit. They can likewise shield you from online tracking by deactivating cookies and scripts that sites try to load.

Open the **Settings** app.

Next, press **Safari**.

Under General, **touch** Content Blockers.

To activate content blockers, flip the switches to the ON position.

Note: The Content Blockers option doesn't appear in Safari's settings until you've installed a third-party content blocker from the App Store.

Temporarily Disable Content Blockers in Safari

Open Safari on your iPhone and go to the site in question.

Next, press the "**aA**" icon in the upper left corner of the screen to uncover the **Website View** menu.

Press Turn Off Content Blockers.

If you only need to disable content blockers for a particular website, tap **Website Settings** in the **Website View** menu, and afterward flip the switch next to **Use Content Blockers** to the OFF position.

Messages

Set a Name and Photo for Your iMessage Profile

Follow the steps below to add your name and picture to your Messaging app.

Launch the **Messages** app.

Tap ••• at the top right side of your screen.

Select Edit Name and Photo.

Then select **Choose Name and Photo** on the next screen.

Input your first and last name, then tap **View More** and choose the photo that you want to use for your profile.

Click **Edit** to choose a picture from your album. Alternatively, select an Animoji from the Animojis displayed.

If you clicked on an Animoji, you will be asked on the next screen to **Select A Pose**. Choose the pose that appeals to you.

Tap **Next** to get to the **Select a Color** screen.

Choose the color that you like.

Tap **Done** and you will be returned to the Profile name screen.

Tap **Continue**.

Tap **Use** if you want to use the picture for both Apple ID and your Contacts. Otherwise, tap **Not Now**.

Tap **Continue**.

Select who should be able to view your name and picture. Tap **Contacts Only** if you want to grant access to all your contacts or tap **Always Ask** if you want to manually select each time you send a message.

Tap **Done**.

Change Your Profile Photo

Launch the **Messages** app.

Tap ••• at the top right side of your screen.

Select Edit Name and Photo.

Tap **Edit**.

Tap All Photos.

Click on the picture you want to use.

Fit the picture into the circle.

Add your filter.

Tap **Done**.

Select Your Initials As Your Profile Picture

Follow the steps below to use your name initials as your profile picture:

Launch the **Messages** app.

Tap ••• at the top right side of your screen.

Select Edit Name and Photo.

Tap **Edit**.

On the next screen, under Suggestions, you will find an image that contains your initials. Select the one that you like.

Choose your preferred color on the following screen.

Tap **Done**.

Sending Messages

Set up your Device for iMessaging

From the Settings app, go to Messages.

Enable iMessages by moving the slider to the right.

Set up Your Device for MMS

From Settings, go to Messages.

Enable MMS Messaging by moving the slider to the right.

Compose and Send iMessage

From the Message icon, click on the New Message option at the top right of the screen.

Under the "To" field, type in the first few letters of the receiver's name.

Select the receiver from the drop-down.

You will see iMessage in the composition box only if the receiver can receive iMessage.

Click on the Text Input Field and type in your message.

Click on the send button beside the composed message.

You will be able to send video clips, pictures, audio, and other effects in your iMessage.

Compose and Send SMS

From the Message icon, click on the new message option at the top right of the screen.

Under the "To" field, type in the first few letters of the receiver's name.

Select the receiver from the drop-down.

Click on the "Text Input Field" and type in your message.

Click on the send button beside the composed message.

Compose and Send SMS with Pictures

From the Message icon, click on the New Message option at the top right of the screen.

Under the "To" field, type in the first few letters of the receiver's name.

Select the receiver from the drop-down.

Click on the Text Input Field and type in your message.

Click the Camera icon on the left side of the composed message.

From Photos, go to the right folder.

Select the picture you want to send.

Click Choose and then Send.

Create New Contacts from Messages on iPhone

Go to the Messages app.

Click on the conversation with the sender whose contact you want to add.

Click on the sender's phone number at the top of the screen, then click on Info.

On the next screen, click on the arrow by the top right side of your screen.

Then click Create New Contact.

Input their name and other details you have on them.

At the top right hand of the screen, click on Done.

Hide Alerts in the Messages App on Your iPhone

Go to the Message app on your iPhone.

Open the conversation you wish to hide the alert.

Click on ⓘ at the upper right corner of the page.

Among the options, one of them is Hide alerts; move the switch to the right to turn on the option (the switch becomes green).

Select Done at the right upper corner of your screen.

FaceTime

FaceTime uses voice and provides a variety of ways to reduce background noise and FaceTime Photos for a visual experience. The network view shows people in your FaceTime call group on a battery of the same size that automatically lights up the speaker. You can also invite everyone to join the FaceTime call using an online connection.

Related messages, photos, and other content delivered to you in messages are displayed in a new section "Share with you" in the same application (available in "Photos," "Safari," "Podcasts," "Apple Music," "Apple News," or Apple TV programs). Many photos published in "Messages" now look like a collage or collection of photos that you can scroll through for easy browsing.

Making a Call

In the Phone application , touch keypad.

Do one of the below:

- **Utilize Another Line:** Touch the line at the top, and pick the other line.
- **Type the Number Using the Keypad:** If there is an error while typing, touch × to delete it.

- **Re-Dial the Last Phone Number:** Touch to see the number of the person you called last, and touch to call that number.
- **Paste a Copied Phone Number:** Long-press the number field above the keypad, and touch Paste.
- **Enter + to Make International Calls:** Press and hold the **0** key till + shows.

Touch the Call icon to start making a call.

Press the Cancel button when you're done.

Re-dial or go back to a recent call
Touch **Recent**, and pick one number to call.

For more info about the call and the caller, touch the More info icon .

Call somebody on your contacts list
In the Phone application, touch Contacts.

Touch the contact, and touch the number you plan on calling.

Call Emergency Numbers When Your iPhone Is Locked

On the password screen, touch the Emergency button.

Call the emergency number (for example, 911 in the US), and touch the Call button.

Reject or Answer Incoming Calls

You can answer, reject, or silence incoming calls.

To Receive a Call:

Do one of the below:

Touch .

If your iPhone is locked, drag the slider.

Muting a call:

Press your iPhone's side button.

To Reject an Incoming Call:

Do one of the below:

Double-press the side button quickly.

Press .

Begin a Conference Call

In GSM calls, you can set up a call and have up to 5 people in the call at once.

In an ongoing call, touch Add call, make another call and touch the Merge call button.

Repeat to add more people to participate in the call.

While on a call, you can do any of the below:

- **Talk Privately with One of the People Participating in the Call:** Touch the More info button ⓘ, and touch the Private Button beside the individual. When you are done with your private conversation, Touch Merge Call to continue the conference call.
- **Add an Incoming Call to the Same Line:** Click on Hold call + Answer, and tap on Merge call.
- **Remove Somebody:** Touch the More Info button ⓘ close to the person, and touch End.

Set Up a Different Ringing Tone for One of Your Contacts

Enter the Contact application .

Pick the contact, touch Edit, touch Ringing tone, and then pick a sound.

Block voice calls, FaceTime calls, and messages, from people
In the phone application, do one of the below:

Click on Favorite, voice mail, or Recent. Touch the More Information button beside the contact or phone number you plan on blocking, scroll down, and touch the block this Caller button.

Touch Contacts, touch the contact you plan on blocking, scroll down, and click the block caller button.

To manage the contacts you blocked, head over to the setting application, tap on the phone, touch Blocked Contacts, and tap on Edit.

Create Contact

Enter the Contact application, Touch +, then enter the contact information.

Look for a Contact

Tap on the search box at the top of your contact list, and write the name, number, address, or other contact details.

Delete a Contact

Head over to the contact card, & tap on the Edit button.

Scroll, and touch Delete.

Share a Contact

Tap on the contact you want to share, touch Share Contacts, and select a way to send the contact info.

Face ID

Utilize the Face ID feature to unlock your iPhone, allow purchases and payments, and gain access to third-party applications by simply looking at your device.

Disable Face ID

Head over to the Settings application, touch Face ID and passcode.

Do one of the below:

- **Deactivate Face ID for some items:** Disable one or more options: iTunes and Application Store, Safari AutoFill, Apple Pay, or iPhone Unlock.
- **Disable Face ID:** Touch Reset Face ID.

Set Up Your E-mail Account

Enter the Setting application, touch Mail, tap on Account, and touch Add Account.

Do any of the below:

- Tap on any email service, like, iCloud or Microsoft Exchange, and then write your e-mail account details.
- Tap on Other, touch the Add Account button and write the needed details to open a new account.

Compose an E-mail

Tap .

Touch inside the e-mail and write what you want.

To edit the format, tap ‹ in the menu bar on top of your keyboard, and clickAa.

You can make changes to the style of the font, the color of the text, use bold or italic, etc.

Add Recipients

Touch the To field and write the receivers' names.

Touch ⊕ to enter the Contacts application and add receivers from your contacts.

If you want to send a copy, click on the CC/BCC box and do any of the below:

- Tap the CC box, and write the names of the individuals you plan on sending a copy to.
- Tap on the BCC field, and write the names of the individuals you do not want other recipients to see.

Reply to an E-mail

Touch the E-mail, touch⇦, and touch Reply.

Write what you want.

Adding an Attachment to Your E-mail

You can add videos, photos, and scanned docs to your e-mail.

Adding a Document
Tap on the e-mail you plan to add the doc and tap on the Expand Tools icon‹ on the format bar on top of your keyboard.

Tap 🗋 on top of your keyboard, and look for the file in the Files application.

In the Files application, touch "Browse" or "Recent" at the lower part of your screen, and tap a folder, location, or file to open it.

Tap the doc to add it to your e-mail.

Scan doc to add to the e-mail.

In the e-mail, Tap where you want to upload the scanned file and tap on the Expand Tool button ‹ on top of your keyboard.

Tap the Docs scanner icon on top of your keyboard.

Arrange your Phone in a way that the doc page shows on your screen—your iPhone would automatically take the page.

To snap the page manually, touch ○ or touch any of the volume buttons. Touch ⚡ to switch the flashlight on or off.

Scan more pages, and touch Save when done.

Deleting an E-mail

There are a lot of ways to delete an e-mail. Do any of the below:

While viewing an e-mail: Click on the delete button at the end of the e-mail.

In the e-mail catalog: Swipe any e-mail to the left, and pick Trash from the menu.

Delete more than one e-mail at once: If you checking out the e-mail list, tap Edit, pick the e-mails you plan on erasing, and tap the Trash button.

Print E-mail

In the e-mail, tap the More action button ↩, and tap on the Print button.

Notes: Utilize the Notes application to write ideas or organize detailed data with checklists, images, weblinks, scanned docs, etc.

Translate App

You can translate text, audio, etc. in the translation app. You can also download languages so that you can translate languages when you are not connected to the internet or when On-Device mode is activated.

Apple's new translation app is a useful app that provides on-site language translation on your device without the need for an internet connection. Simply enter a word or phrase, select the language you want to translate, and the phrase will appear. You can write the statement or say it verbally; the app will write it down and reply to you. For bilingual conversations, put your phone in landscape mode and let Siri recognize each language. Supports up to 11 languages.

Access App Clips

App Clips allows you to use enhanced versions of applications without having to download full versions and store them in the application library, where they will not take up valuable space on the home screen. You can access the clips in the app from a variety of sources, such as clicking a link on the web page, message, maps, or Safari QR code.

Add audition to the Control Center with Settings> Control Center. As you playback, your recording, the Listen icon monitors the sound output to indicate safe or unsafe sound levels. If you press and hold the icon, you will get the decibel level where anything above 80 decibels is considered unsafe.

Translate your voice or text

In the Translate application, touch translation, choose the language to translate it, and do any of the below:

- Touch the **Enter text** button, write the phrase and touch the **Go** button.
- Touch the Listen button, then say the phrase.
- After translating, do any of the below:
- Play what you translated: Press the Play key.
- Save what you have translated as a Favorite: Click the Favorite button.
- Find a word in the dictionary: Touch , then touch a word to see its meaning.
- Show what has been translated to others: Press the Full-Screen button .

Tip: Swipe the translation down to see your history.

Translate a Conversation

Your device would translate your conversation and would display it in text bubbles from the two sides of the conversation. You can also download languages for offline translating.

Touch Conversation

Click the Listen button, then speak in one of two languages.

Tip: A conversation can be translated without you pressing the Mic button before everyone speaks. Touch the More Options icon, touch the Auto Translate button, and touch to begin the conversation. Your iPhone would automatically notice when you begin to speak and when you stop talking.

When you're discussing face-to-face, touch, and touch the **Face to Face** icon so that everybody can see their side.

Download languages for offline translating.

Enter the Setting application, and touch Translate.

Click on Downloaded Language, and touch beside the languages you plan on downloading.

Enable On-Device mode.

Connect Your Phone with Other Devices

Share your internet connection.

Utilize the Personal Hotspot feature to share your internet connections with other devices.

To Set Up Personal Hotspot

Enter the Settings application, touch Cellular, touch Personal hotspot, and activate **Allow Others to Enter**.

You can adjust the following settings:

- **Make Changes to your Wi-Fi Passcode:** Enter the Settings application, touch Cellular, tap on Personal Hotspot > Wi-Fi Passcode.
- **Make Changes to Your Domain Name:** Enter the Settings application, touch General, tap on About, and touch Name.
- **Deactivate the Personal Hotspot:** Enter the Settings application, touch Cellular, tap on Personal Hotspot, and deactivate **Allow Others to Enter**.

To connect to your Personal hotspot, enter the Setting application on the other device, touch Wi-Fi, and select your device from the catalog of networks.

After connecting the other device, a blue band would show at the top of your screen.

Hand off tasks between your Apple devices.

With the Handoff feature, you can start something on your iPhone and finish it on your MacBook, iPad, or any other Apple device. For instance, you can start replying to an e-mail on your iPad and finish it on your MacBook. You can use Handoff in a lot of Apple applications, such as Safari, Contacts and Calendar. Some third-party apps might work with the Handoff feature.

Social Media Apps

Launch the App Store from your iPhone's home screen.

At the bottom of the screen, tap Search.

Enter the name of the app example. “WhatsApp, Facebook, Twitter, Instagram" in the text box.

Tap the download icon to the right.

Once downloaded, tap "Open."

Managing Files

Zip and Unzip Files

Zipping and unzipping files is now easy with iOS and requires no third-party app.

Zip Files

You can easily compress files by using the Files app on your iPhone.

Launch the Files app and go to the folder containing the files you intend to zip.

Press **Select** on the top right.

Mark the files you'd like to zip.

Press **More** on the bottom right and press **Compress**.

An Archive.zip file will display within that same folder and contain the files you zipped.

Unzip Files

You can see how easy it is to zip files on iOS, and unzipping them is even simpler.

Launch the Files app and go to the zipped file you want to unzip.

Click on the file.

The file will then unzip and turn into a folder within that same folder. Press to access the contents.

Scan Documents from the Files App

Launch the Files app on your iPhone.

Press the **Browse** tab at the bottom of the Files app.

Press the **More** button (three-dot icon) at the top of the display.

Press Scan Documents.

Place your document in the viewer and touch the Capture button.

You can decide to drag the corners to modify it, touch to **Retake** or touch to **Keep Scan**.

If there's another page you want to scan, you can just capture it on the next screen.

When you're done scanning, press "**Save**."

Select a location for your scan and press "**Save**."

You can also scan a document within a location like iCloud Drive or On My iPhone in the Files app.

Tap the **More** button (three-dot icon) at the upper left and then follow the same steps above.

Save and Share Webpage as a PDF

Launch the Safari app on your iPhone and visit any web page of your choice and let the page get loaded completely, else, it will not be able to save the full page as PDF later on.

Now press and hold the Home button and Side button at once to capture a screenshot on your iPhone in Safari.

You can now see the preview of the screenshot taken in the bottom left corner, tap on the screenshot and then press the **Full Page** option that's available in the right top corner.

Next, tap on "**Done**" and then select **Save PDF to Files**" option.

Select any of the folders from "**On My iPhone**" or "**iCloud Drive**". If the desired folder isn't available, create one and then press "**Save**." This will save your PDF.

If you want to share the PDF via e-mail or iMessage, after the third step, tap on the share button option available in the top right corner.

Once done, select iMessage/Mail or any other platform, enter the recipient, and press "**Send**."

CHAPTER 9:

Efficiency and Adequacy Apps

Trying to find a place, read a map, or get directions? Apple's iOS has it all! This article will go over the many utilities and maps apps that are available on the iPhone. Read on for more information!

---

iOS Apps: Utilities and Maps: What They Are and How They Work

All of the latest iOS updates offer amazing features for managing your device.

What's New in iPhone OS 3.0?

iPhone OS 3.0 is packed with amazing features that improve your experience with the device. iPhone OS updates fix bugs and boost security and usability. The latest update for iPhone and iPod Touch (version 3.0) was released on June 17th, 2009, and it comes packed with a ton of new features that improve the usability and security of your device.

The Apple App Store, also known as the iTunes App Store, is a digital distribution platform developed by Apple Inc. that distributes media content and software applications to desktop computers and portable devices running the macOS, iOS, and Windows operating systems. The store was launched on April 28th, 2008 with an initial 500 applications available for download. Subsequently, it was expanded on September 9th, 2008 with an additional 250 titles. It was opened to the public as part of a software update of Mac OS X 10.5 "Leopard." It was also accessible during early public beta testing of Mac OS X 10.6 "Snow Leopard."

It is designed to be the primary sales channel for all applications developed by Apple and its partners. The App Store is the only store that accepts apps built with the iOS SDK (software development kit). It allows users to browse and download free and paid applications, view reviews and previews of new apps, view updated news, and view recommendations from Apple's App Store staff. The app store has over 2 million apps on its database including games such as Angry Birds, photo editors such as iPhoto or GarageBand, music players such as iTunes or iConcert, and third-party games like Chess Titans 2 or Sweet Escape 2. The marketplace generates more than 1 billion app downloads per month.

The map app that comes pre-installed on iOS devices (such as the iPhone, iPod touch, and iPad) is a vector-based, turn-by-turn navigation tool using GPS. This app uses Apple's new vector-based rendering engine and is highly accurate in terms of global positioning. Some examples of its features include:

Apple Maps supports turn-by-turn directions with real-time traffic information, voice guidance prompts, and multiple routes (which includes public transportation). With the iOS 5 upgrade, you may also view the map in 3D mode. You may see placemarks as well as business details.

Using Siri, you can ask for your current location or how to get somewhere. Asking for directions will launch the map app and display the route. Siri will also provide a list of contacts that you can call to navigate to that location.

You can view a list of places near your current position, along with detailed information about each place (from Yelp) and reviews from other users to help guide your decision as to where to go. You can also create a new contact for that business, find out more about the business (from Facebook), look up directions or call the business from within the map app itself.

Notes

Notes are available on iPhone OS 3.0. It allows users to create text notes or sketches using their finger or stylus, as well as e-mailing or printing via AirPrint. Notes are synced to the cloud, allowing you to access them across numerous platforms.

The Notes app also supports saving a note via Siri (if the user is connected to a network). "Take a note," Siri can be asked. The command will save it in the Notes app, and it will have a link that takes you directly to that note, as well as any other information entered in the "Note" itself.

You will have a set of pre-defined templates for your notes. You can add photos, clip art, and other content to the top menu area of your note, as well as text.

Currently, there are three "Notes" included with your device:

Sketchbook: Create sketch or doodle drawings in a digital sketchbook.

Notepad: Add text and other items in a simple text editor.

Reminders: Add pre-set reminders to lists (this applies to the information entered). This is what you would expect from an application that allows you to create reminders.

Notes can be synced with the Mac OS X version of Notes via iTunes or iCloud. iCal and Mail can sync your reminders information to and from your iPhone, however, they are not capable of doing so with text notes.

You can print a note via AirPrint; this also saves a digital copy onto your device. Select "Email" in the top right corner of the note editor screen to email a note to yourself or someone else. This will allow you to attach a note to an email or email it as a PDF document. You can attach photos or movies to an e-mail if you have them on your device.

The Notes app is the default application for saving and managing text notes on Apple's iOS operating system. On iOS 4, it was moved from its original location on the third page of applications to its new position on the first page of applications. It is available for iPhone, iPod Touch, and iPad devices running iOS 4 and newer.

Although the Notes app was available since iPhone OS 1.2, it gained popularity after it received an overhaul in iPhone OS 3.0 with a new user interface, a feature that allows you to checkmark items off lists such as grocery lists ("Reminders"), pictures and more. It also has file attachments capability through email or AirPrint (which will save as a PDF file).

Reminders

The Reminders app is an application that is available on Apple's iOS. Reminders were released in January 2011.

The app was updated with the release of iOS 5, which included a number of improvements such as:

Search: Users can search their reminders by title, list, or the due date (this includes searching via voice recognition from Siri).

Bill Reminder: Users are able to mark bill reminders as paid. It will then display on the "All" tab.

Location-based reminders – Create Reminders for places you go often, or when entering/leaving a particular location (iOS 5 only).

Reminder Lists: Create multiple lists and switch between them to organize your groups of reminders.

Reminders can be downloaded and viewed offline when connected to a Wi-Fi network. Reminders can be synced with the Mac OS X version of Reminders via iTunes or iCloud. Apple released an application entitled "Passbook" in iOS 6, which is a unified location for storing digital versions of gift cards and loyalty cards. It also includes boarding passes, event tickets, and coupons. Simply place your phone over a card reader (such as those used in supermarket stores) and the code on the card or pass will be scanned and added to Passbook. This app was developed for convenience, so users don't have to carry around a bunch of cards in their wallets or purse. iOS 7 introduced transit tickets and mobile boarding passes, which can be synced to the Wallet app on your iPhone. These tickets can come from email, the web, or from a third-party application such as Air Canada's Aeroplan. Apple has introduced an iPhone software named "Wallet" for iOS 6 and iOS 7 users that allows them to access their gift cards more easily. Instead of going into Passbook, it brings up a full-screen view of all gift cards in Wallet so that you don't have to go through as many steps to access your cards. The App Store is a digital distribution platform where Apple Inc. helps developers sell their apps and digital assets.

Events on the Calendar

The app is used to read events stored on the device's calendar. It also syncs with MobileMe and soon-to-be iCloud or syncs directly with online calendars via the Lightning connector (which will allow users to dock their iPhone to a computer). In addition, it can be accessed with Microsoft Outlook and Google Calendar. The calendar also supports accessibility features such as VoiceOver technology.

Contacts on the Phone

Here you can keep a list of contacts on your phone's address book through your cellular carrier and iPhone. The default contacts application was first introduced in iOS 2.0 and replaced the phone book from iOS 1.x which was stored in contacts.app in iTunes 9/8/7 for Mac OS X.

Calendar on the Phone

The calendar app enables access to your calendars stored on the device's local storage. Contacts, Mail, Notes, and Reminders can also be synced via iCal. The default calendar app was first introduced in iOS 2.0 and was later updated in iOS 5, which added a variety of new features such as a month view when swiping left and right (which was previously available on Mac OS X).

Music Store

The Music app stores music that is purchased from Apple's iTunes Store (or from other retailers who support the store), as well as music that is downloaded from iCloud or other sources through iTunes Match. iOS 7 introduced an in-line, iMessage-Esque messaging feature within the Music app. Users can send messages to other users, who are also in their contact book (or by default, anyone who has already accepted a friend request). These messages can be audio or video and will appear with a timer before being delivered. This feature only works between two devices that are both running iOS 7 and Apple's iCloud service.

Pictures on the iPhone

Pictures on the iPhone are stored locally by default so that they do not need to be transferred back to the computer or any cloud services such as iCloud.

They can also be synced with iCloud or stored on your computer through iTunes.

The iOS 7 app also allows users to access the photos and videos they have taken on their iPhone using a built-in search feature that works across all of the photos on the device and from the Photo Stream.

GoodReads

It was released in 2011, allowing users to read reviews for books and track them with ease.

Users may browse for new books and download free e-books from the iBookstore using the app, which is accessible for both iPhone and iPad devices. The app supports bookmarks, highlighting, next page links, and syncing with a GoodReads account.

RunKeeper

It is a web application that allows runners to post live videos of their runs to a website where others may watch them. The app has live tracking of runs, and running statistics (pace, speed, time). It also provides users with an interactive map of the location of their runs.

- App for Clock
- App for Calculator
- App for Weather

Safari

It is a web browser used on iPhones and iPads. The app supports HTML5 and JavaScript, as well as functions as a PDF viewer. It can be accessed both with iOS 6 and iOS 7.0 or later. Safari is integrated with the App Store, allowing users to purchase applications directly from the Safari web browser.

iBooks

iBooks (formerly iBookstore) is the digital bookstore application that was released on September 10, 2011, for the iPhone, iPad, and iPod Touch; making it the first universal application for these devices that allows owners to read e-books on their mobile phones or tablet devices without having to download an app akin to an e-Reader.

The iBookstore also allows users to purchase and download e-books on their iOS devices as well as rent them and read them.

App Store

It is an app that allows users to browse, purchase and download applications. It was released with the iPhone in 2008.

Camera

It is a built-in application on the iPhone that lets users take pictures and record videos. It was originally standard on all iPhones since its introduction in 2007, but starting with the iPhone 5s it became a paid download for new devices.

The app also has Siri integration which allows for voice commands for taking pictures or recording videos by saying "Cheese" or "Smile," among other things, although not everything will work.

How to Use a Stopwatch or Timer

The App Store is a digital distribution platform where developers can sell their apps and digital media with the assistance of Apple Inc. The app allows users to browse what's available on the App Store or look for new developments. The App Store enables users to download paid apps and in-app purchases (apps that are purchased within other apps) that are developed by third-party developers, such as themselves or from independent software developer labs (ISDLs). The App Store also enables in-app purchasing of virtual currencies in some games, such as "IAPgalore" (developed by In-App Purchasing Labs).

The Apple TV has a dedicated app store which was released with the 3rd generation Apple TV on March 21, 2012.

iCloud is a cloud-based service that can be used on Mac OS X, Windows, and iOS. It allows the user to store any type of data, including files, websites, contacts, and calendars on remote servers over the Internet. It was first released in October 2011.

iOS 8 added a new method of controlling (and receiving notifications) called “System Notifications.” This feature also introduces a new set of larger icons that are used to dismiss notifications as well as interact with them using App Switcher. This app is only available for iOS 8 on devices with 3D Touch—the feature itself requires a Force Touch display and a compatible device to function in this manner, such as iPhone 6s Plus or the Apple Watch Edition.

Many of the apps were used with earlier versions of iOS, but have been updated to use the new features found in iOS 7.

Third-party applications are not natively coded to run specifically on iOS. Instead, they are compiled for iOS by either their developers or distributor; Apple uses software development kits and adheres to guidelines imposed by Apple on distribution through the App Store. This way of developing for specific platforms contrasts with that of Android where software is usually compiled for a platform using an SDK provided by the manufacturer of a smartphone or tablet.

The primary purpose of the App Store is to provide an interface for user purchase of application programs for the iPhone, iPad, and iPod Touch. Apps can also be bought and downloaded from other websites that allow iOS users to browse for and purchase non-native apps.

Many applications are free, but others have a price tag attached, ranging from free (such as weather stations) to $99.00 or more (specialized medical software). Apple has a policy that all apps must be functional without the need for additional items (such as a camera) other than those provided with the device.

All apps submitted and approved on or after October 6, 2011, must be compatible with the iPhone 5's 4-inch display.

On November 28, 2016, following the release of iOS 10.2, Apple announced that iOS apps developed using Objective-C or Swift cannot be distributed through the App Store to devices running earlier versions of iOS; in particular, iPhone 4 and iPad 2 are excluded from running applications built with these languages. Apps can still be developed in either language prior to submitting them for Apple's approval.

Apple's in-app purchase and advertising policies have been criticized by users and developers. The App Store is one of the more profitable sources of revenue for Apple and saw revenue increase by approximately 50% during the second quarter of 2012 compared to 2011.

Apple has been criticized for having a 30% transaction tax on all transactions completed through the App Store, with no discretion in implementation or conditions applied. On July 12, 2012, Apple removed the transaction tax from all iOS apps with in-app purchases after criticism it would have a disproportionate effect on smaller developers using an estimated 70% of iPhone app revenues, while larger companies could afford to pay the tax.

On January 17, 2012, Apple instituted a radar gun app policy requiring developers to inform users about the data transmitted back to them for advertising purposes.

The controversy that ensued was due in part to Apple's much-criticized handling of existing privacy concerns coupled with the fact that this was seen as an end-run around giving users the option to opt-out of data collection (i.e., by not allowing such apps into their phones in the first place).

In June 2013, it was revealed that some free apps are accessing user data such as consenting adults' contact details, including names, email addresses, and phone numbers listed under FaceTime and iMessage.

As a result, Apple has revised its rules to make it illegal for apps to "collect information about which other apps are installed on a user's device for the purposes of analytics or advertising/marketing."

In April 2016, Apple announced that all iOS apps submitted to the App Store must support IPv6 by June 2016 and comply with new IPv6-only guidelines. This is because an increasing number of networks and Internet routers have begun supporting IPv6 but older versions of Apple iOS devices only support IPv4 (therefore only allowing connections to those networks via carrier-grade NAT). This causes devices running unsupported versions of iOS to be unable to access services from any network using IPv6 as well as services not using a Carrier Grade NAT.

CHAPTER 10:

Wellness

What to Do First

You will need to provide your Medical Details in these three profile sections:

1. **Health Profile:** It contains your Name, Contact, Date of Birth, Sex, Blood Type, Fitzpatrick Skin Type, and Wheelchair.
2. **Medical ID:** It shows the same emergency information on your iPhone or Apple Watch which also contains activation of emergency when your iPhone is Locked, Date of Birth (DOB), Medical Conditions, Medical Notes, Allergies and Reactions, Medication, Blood Type, Organ Donor, Weight, Height, and Emergency Contacts.
3. **Organ Donation:** This will enable you to register as an Organ donor through the **sign-up with Donate Life**. It comprises the following data of yours; Names, DOB, Last 4 Social Security Number (SSN), Email, Address, ZIP, and Sex.

How You Can Set Up Medical ID

Home Screen: Hit on the **Health App** icon.

Today Page: Hit on **Medical ID icon** at the right bottom of the screen *

Medical ID: Hit on **Create Medical ID** bar.

Medical ID *

Hit on **Show When Locked** activation slider to enable access emergency when your iPhone is locked.

Hit on **Add Photo** to upload your photo on the page.

Hit on **Add Date of Birth:** You will see the date suggested at the bottom of the screen. Scroll up until you will get to your Month/Day/Year and hit on it to select.

Hit on **Medical Condition** to type your health discomfort (e.g., Herpes, stomachache, arthritis, etc.).

Hit on **Medical Notes** to write medical history.

Hit on **Allergies and Reactions** to write the signs and symptoms in the text field. I do not have any reaction you may write "Nil/None."

Hit on **Medication** to write the doctor's prescription in the text field. If you do not have, you may type "Nil/None."

Hit on **Add Blood Type** to select.

Hit on **Organ Donor** to select "No, Yes, or Not Yet".

Hit on **Weight** to select your body weight.

Hit on **Height** to select your height.

Hit on **Add Emergency Contact** to select your trusted loved ones' contact from your iPhone contact list.

Hit on **Done** at the right top angle of the screen.

How You Can Set Up Your Health Profile

Home Screen: Hit on the **Health App** icon.

Health: Hit on **Summary Tab**.

Summary: Hit on your **Profile Picture** at the top right side of the page.

Your Name Page: Hit on **Health Profile**

Health Profile:

- Hit on **Edit** at the top right angle of the page.
- Provide your age, weight, and height
- Hit on **Done** at the top right angle of the page

To Confirm How The Emergency Settings Work

Press the **Power** Button to sleep and re-press the button to wake the iPhone.

Press **Home Button** to see the **lock screen** with the **Emergency** at the bottom left of the screen.

Hit on the **Emergency**

Emergency Call: If you must access Medical ID, then hit the **Medical ID** at the bottom left of the screen. All the information you have provided during the setup will be shown.

At the bottom of the Medical ID page, you will see the **Continue bar**. Hit on the **Continue** bar to complete the registration. Hit on **Complete Registration with Donate later** bar.

Thank You: Hit on the **Done** bar.

How You Can Setup Your Health Data

Home Screen: Hit on the **Health App** icon.

Dashboard: You will see your **Dashboard**. There are period titles which are **Day | Week | Month | Year**. You can select any of the periods that you will be tracking your health data. To start with, select **Day** by tapping on it.

Hit on **Health Data** at the second left bottom of the screen to add different health categories such as **Body Weight**, **Active Calories**, **Heart Record**, **Heart**, **Reproductive Health**, **Body Temperature...** and more others.

For instance:

Hit on **Active Calories**.

Active Calories: Hit on **Add Data Point**.

Add Data:

Type your daily preferred Calories figure like 250.

Hit on **Add** at the top right angle of the screen.

You will see that the number of calories has been increased from the recommended calories per day. But if you are not having calories on the displayed table, you could go ahead and add your daily calories. Your iPhone will strictly use your daily recommended calories to track your daily calories used through your daily activities.

For the tracking to show on your Dashboard, then you have to activate **Show on Dashboard** slider.

Now as you are moving around, the number of your steps, distance covered in Kilometers (Km), and Calories spent (kcal) will be tracked.

How to Add More of Your Health History

Home Screen: Hit on the **Health App** icon.

Health Data:

Hit on the **Browse** tab.

Hit on a **Category** like Activity.

Select **Subcategory** like Steps.

Hit on **Add Data** at the top right angle of the screen.

Type the **Date**, **Time**, and **Data** for the Activity.

Hit on **Add** at the top right angle of the screen.

How You Can Track All Your Favorite Categories with Your iPhone

Home Screen: Hit on the **Health App** icon.

Today Summary: Hit on **Summary Tab.**

Summary: Hit on **Edit** at the top right angle of the screen page.

Edit Favorite:

Hit on each desirable Star to add more **Categories**  .

As soon as you have completely selected all your preferred **Favorite Categories**, hit on **Done** at the top right side of the screen.

Hint: When you hit on each star, it will change from a framed star to a full blue star .

How You Can Check Your Health Data

Home Screen: Hit on the **Health App** icon.

Health Data: Hit on the **Category** like **Body Measurement**.

Body Measurement: Select your **Subcategory** like **Body Mass Index**.

Body Mass Index:

Put "On" the **Add to Favorite** activator. Hit on **Show All Data** to see all the data of the tracked activity. After you have confirmed what you want to know then you can back the page at the top left side of the page.

How Organ Donation Could Be Setup on Your iPhone

Home Screen: Hit on the **Health App** icon.

Health: Hit on **Summary Tab**.

Summary: Hit on your **Profile Picture** at the top right side of the page.

Your Name Page: Select **Organ Donor**.

Organ Donor: Hit on the **Sign Up With Donate Life** bar to sign up.

Registration

All your profile details will display, what you only need is to enter the last four-digit number of your **Social Security Number** (SSN).

Hit on the **Continue** button below to complete your registration.

Hit on **Complete Registration with Donate Life** at the bottom center of the screen.

Thank You: Hit on the **Done** button below.

Your complete registration with Donate Life will show up under your Medical ID profile as **Organ Donor-Donate Life**.

How You Can Fix Failure of Health Tracking on Your iPhone SE

For your health tracking walking steps to be perfectly working, you will need to connect your Apple Watch with your iPhone.

If you are having a challenge of not seeing your walking **steps** being tracked by the Health App on your iPhone SE then take the following steps:

Home Screen: Hit on the **Health App** icon.

Health: Hit on **Summary Tab**.

Summary: Hit on your **Profile Picture** at the top right side of the page.

Your Name Page: Under **Privacy** hit on **Devices.**

Devices: Select your **Apple Watch.**

Your Watch Privacy: Hit on **Privacy Settings** and activate **Fitness Tracking** by switching on the activator.

How You Can Make Family Sharing Working

Home Screen: Hit on the **Settings App** icon.

Settings: Go to your **Profile Name** at the top.

Apple ID: Select Set Up Family Sharing.

Family Sharing: Hit on the "**Get Started**" button.

Get Started: Select **Location Sharing** or any other feature you want to share. There might be a slightly different method of setup in various other features of family sharing available.

Share Your Location With Your Family: Hit on **Share Your Location** button at the bottom center of the screen. If you are not ready to share your location you can select the **Not Now** option below.

Invite Your Family: Hit on the Invite Family Members bar at the bottom center.

Perfect Ways of Using Health App to Track Menstrual Cycle on Your iPhone

The use of Health apps to track women's menstrual cycle can only be studied on your iPhone SE because it is running compatible with iOS 13, and it can also be observed on an Apple Watch running watchOS 6.

However, it is very important you use Apple Watch with your iPhone SE running iOS 13 to benefit more from the advanced innovativeness and efficiency loaded in the Health App.

How You Can Set Up Every Month Menstrual Cycle on Your iPhone

Many females that are having a monthly menstrual cycle with the irregularity of period flow and starting date, or inability to determine the ovulation period will be able to calculate their menstrual interval, supposed ovulation period... and many others.

Home Screen: Hit on the **Health App** icon.

Search: Hit on the **Browse** tab at the bottom right of the screen .

Browse:

Select **Cycle Tracking** among the **Health Categories** .
Hit on the "**Get Started**" button.

Cycle Tracking: Hit on **Options**

Options:

You will scroll down to select **Period Length** and swipe up to select your period duration, e.g., 2 days, 3 days, 4 days, 5 days, etc. The period length is the number of days you experience menstrual flow. For example, if you see the menstrual flow from August 2 to 5 is equal to 4 days.

Select **Cycle Length** and swipe up to select the appropriate days before the next menstrual cycle e.g. 25 days or 26 days, etc.

Hint: If you are having some pre or post symptoms that come with your cycle you may activate all the relevant suggested contents above **Your Cycle Log**. Some of them are:

1. Spotting
2. Basal Body Temperature
3. Cervical Mucus Quality
4. Ovulation Test Result
5. Sexual Activity
6. Symptoms... and others.

How You Can See Your Cycle Timeline

Home Screen: Hit on the Health App icon.

Search: Hit on the Browse tab at the bottom right of the screen.

Browse: Select Cycle Tracking among the Health Categories

The timeline will appear as:

- **Solid Circles:** Number of days you recorded for your menstrual period.
- **Purple Dots:** The number of days you recorded for experiencing symptoms
- **Light Red Circle:** This is your next menstrual period Prediction.
- **For You to Hide or Display Predicted Period Days**: Select Option and activate the Period Prediction switch.
- **Light Blue Days:** These predict your possible Fertility Window. You should not use it to guide yourself for birth control.
- **For You to Hide or Display Infertility Window:** Select Option and activate the Fertility Prediction switch.

How You Can Know Your Possible Next Menstrual Cycle

Home Screen: Hit on the Health App icon.

Search: Hit on the Browse tab at the bottom right of the screen. 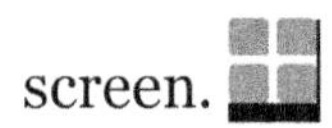

Browse: Select Cycle Tracking among the Health Categories.

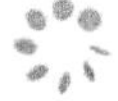

Cycle Tracking:

Scroll down to select Prediction under Cycle Log. This will enable you to know your next menstrual cycle.

If you are unable to see the evaluation: Select Show All before the prediction option.

Scroll down to select Statistics. This will enable you to see all your previous menstrual periods and predictive cycle length.

On Apple Watch: For You to Determine Date 

Launch the Cycle Track App.

Scroll down to select Period Prediction/Last Menstrual Period.

How You Can Track All Your Cycle Symptoms in Health App
Home Screen: Hit on the Health App icon.

Search: Hit on the Browse tab at the bottom right of the screen.

Browse: Select Cycle Tracking among the Health Categories.

Cycle Tracking: Hit on Options in front of Cycle Log.

Options: Activate the Symptoms activator to access all the possible symptoms before/during/after the menstrual cycle. All the symptoms are listed below:

- Abdominal Cramp
- Acne
- Appetite
- Bloating
- Breast Tenderness

- Constipation
- Diarrhea
- Headache
- Hot Flashes
- Lower Back Pain
- Mood Changes
- Nausea
- Ovulation Pain
- Tiredness
- Sleep Changes

You can tap on any of the symptoms you are experiencing before or during or after the menstrual cycle to log your symptoms with Cycle Tracking.

How You Can Record Menstrual Cycle Symptoms on Your Apple Watch

To launch Apple tray by pressing Digital Crown.

Select the Cycle Tracking App icon.

Hit on Symptoms.

Scroll through the various available symptoms and hit on various symptoms lists you regularly experience before or during or after the menstrual cycle.

Hit on Done.

How to Predict Menstrual Cycle, Fertility, and Notification through Cycle Tracking

Home Screen: Hit on the Health App icon.

Search: Hit on the Browse tab at the bottom right of the screen.

Browse: Select Cycle Tracking among the Health Categories.

Cycle Tracking: Hit on Options.**Options:** Hit on the activator of the below features to turn them "On."

Activate Period Prediction. 

Activate the Period Notification. 

Activate the Fertility Predictions. 

Activate Fertility Notifications.

How You Can Record a Period Flow Level Through Cycle Tracking on Your Apple Watch

To launch the Apple tray, press Digital Crown.

Select the Cycle Tracking App icon.

In the everyday tracker above the data, Summary hit on the Day. Below Menstrual Unit, hit on Period to record your menstrual flow. Select your period flow level by tapping on any of these options below:

- Light
- Medium
- Heavy

Hit on Done.

How You Can Further Record Your Sexual Activity in Cycle Tracking

You can use this to track when you last had sex with your spouse.

Options:

Hit on Sexual Activity.

On the same page hit on Sexual Activity and select Had Sex.

Select Not Used or Used to remind you of your sexual protection.

How You Can Record Your Sexual Activity in Cycle Tracking on Your Apple Watch

On Apple Watch

To launch the Apple tray, press Digital Crown.

Select the Cycle Tracking App icon.

Hit on Sexual Activity.

If you had sexual intercourse, then hit on Had Sex.

For protection confirmation, select Not Used or Used.

How You Can Remove All Cycle Tracking in Health App on Your iPhone.

Home Screen: Hit on the Health App icon.

Search: Hit on the Browse tab at the bottom right of the screen.

Browse: Select Cycle Tracking among the Health Categories.

Cycle Tracking: Scroll down to select View Cycle Tracking Items.

View Cycle Tracking Items: Hit on every Category Log you wanted to remove.

Scroll down to the bottom region to hit on Show All Data.

At the top right angle of the screen hit on Edit.

Hit on the Remove icon beside the data you wanted to remove.

Hit on Delete.

Hit on Done at the top right angle of the screen.

Easy Ways of Knowing the Data Source from Several Sources

Health App always makes use of the same data from different sources like Apple Watch, iPhone, and iPod. If you are using several Apple devices that are connected with iPhone XR, it very important as a user to specifically know which of the Apple device is responsible for the data making.

You can set your iPhone to be the major source that Health App should be used for data generation, instead of using all available Bluetooth devices. This method is called the **Prioritization of Device**.

Home Screen: Hit on the **Health App** to launch the page.

Summary: Hit on the **Browse** icon tab.

Browse: Hit on a **Category** such as **Activity**.

Activity: Select the **Subcategory** such as Steps or Walk & Running or any other one.

Hint: But if you are unable to see the subcategory you are looking for, then you can scroll up to access the Searching field tool at the top of the page. Hit the field and enter the name of the **Subcategory** and it will appear.

Steps: Scroll down to select **Data Sources and Access**.

Hint: You will see all the Sources that are responsible for the make of the Steps' Data will be enlisted.

How You Can Access All the General Sources for the Health App

Home Screen: Hit on the Health App to launch the page.

Summary: Hit on Profile Picture at the top right of the page.

Your Profile Name Page: Move down the page to Privacy and hit on Apps or Devices.

You will see all the Sources that enable Health App to generate those Data.

How You Can Assign Sources for Data

This will enable you to prioritize sources for the making of a particular subcategory data in the Health App.

Home Screen: Hit on the Health App to launch the page.

Summary: Hit on the Browse icon tab.

Browse: Hit on a Category such as Activity

Activity: Select the Subcategory

Subcategory: You will scroll down the page and select Data Source & Access.

Data Source & Access: Hit on Edit

Touch and hold down the Change Order button at the front of the Data Source.

You can either drag up or down the Data Source on the lists.

If you do not want a Data Source to generate data for the subcategory: Deactivate the activation switch.

CHAPTER 11:

Introduction to Siri

Apple recently announced the new iPhone, coming out this fall. The new iPhone is expected to cost upwards of $1,000.

Despite its premium price tag, the phone comes with a variety of features that many other smartphones don't have. For example, it has wireless charging technology and an upgraded camera system. But one feature stands out from all others—the voice assistant SIRI. This virtual helper will allow you to control your phone hands-free by simply telling it what you want it to do for you—without having to touch your phone at all!

This will go over everything about this exciting new addition to Apple's world-famous smartphone lineup. If you're paying attention, you'll come away with a better understanding of how SIRI works and why Apple's voice assistant is so revolutionary.

SIRI is a virtual assistant that will work on any Apple device with an operating system that is currently on iOS 11. This includes anything from the newest iPhone to older models including the iPad, iPod touch, and Apple Watch. SIRI will be available to download onto your iPhone, iPad, or iPod touch either through the App Store or by logging into your Apple ID on its website at http://www.apple.com/ios/siri/. SIRI will also be integrated into your watch devices and is expected to have a full suite of functionality available as well as support for third-party applications.

What Is Siri?

SIRI is a combination of Siri and Search Logic. Siri is Apple's voice assistant while Search Logic is its search engine. These two separate products were combined into one product that better works with Apple's voice assistant, resulting in a product that is more powerful, more efficient, and easier to use. SIRI will be on all of the latest versions of Apple's devices including the iPhone X and the new iPad Pro.

Why Is Siri So Unique?

SIRI is unique for many reasons. One is because it combines a search engine with an AI assistant into one package. The other is because of its advanced neural network. This allows it to recognize voice inflections and can understand questions better than any other AI assistant currently on the market.

What Is a Neural Network?

A neural network is a computer that works like a human brain. It collects information and processes it in ways that are similar to how the human brain functions, resulting in more natural human-computer interactions. This makes the virtual assistant experience much easier to use, resulting in what Apple CEO Tim Cook called an "intelligent digital assistant."

However, unlike Siri, SIRI will be able to complete tasks related to more than just work information. For example, SIRI can complete tasks that are related to health, education, and even entertainment.

Why Is the Neural Network So Advanced?

Finding out how to get the greatest results from AI helpers took a lot of time, money, and effort. Apple was one of the companies leading the way in this research by creating Siri with a large neural network powered by 5 GB of RAM onboard for more advanced processes. This has allowed SIRI to be more intelligent than Siri on Apple devices from previous versions. Therefore, it is able to handle more complex tasks and respond better than its competitors.

How to Make Siri Work on Your iPhone

Siri is not available for every single iPhone. You can only get it on the latest model, the iPhone 6S and higher. To use SIRI on your iPhone, you'll have to have iOS 11 installed. This means that all of your Apple devices will have to be updated before SIRI can work on them.

Once you've made sure all of your Apple devices are up to date with iOS 11 and SIRI has been activated, you'll need to put in a passcode on all of the devices because they will be interacting back and forth very closely when using Siri.

Note: If you don't have a passcode, Siri can't work on your device.

To use SIRI on your iPhone or iPad [or the Apple Watch], you'll need to open the correct App Store application. This can only be done on iOS 11 devices of course. First, you'll need to log in to your Apple ID and search for "Siri."

With this done, tap on "Search." If it doesn't tell you that it's installed properly, then tap "Search." On this screen, there are a variety of options including searching for movies, TV shows, music, and apps. Tap the option that best describes what you want and search for it. After that, tap the app or movie and choose the option to "Buy."

After you've bought something, it'll show up in your App Store. After that, you can open it whenever you like by tapping the app or movie on the home screen of your iPhone.

This is how you can use SIRI on an Apple device. It's very simple and easy to do with this guide! Once you have SIRI installed on your devices, there are a few things that you'll be able to do with it such as set reminders and check facts.

Siri Supported Features

Siri is a very impressive AI assistant. It can do a lot of things even outside of work-related tasks. For instance, if you are trying to brush your teeth and your phone rings, you can ask Siri to check who's calling using the "What's Voicemails" feature. You can also set reminders that Siri will check for you regularly on your iPhone or iPad. SIRI will also be able to answer basic questions such as "Who won the Super Bowl?" or "How much does the flight from NY to London cost?" But it's not just about work-related questions. You can also ask Siri to play music or sports updates or tell you a joke. SIRI will be able to do this, along with a lot more information, on your iPhone and iPad as well as your Apple Watch.

Siri can be taught to recognize your voice.

This is a very interesting feature. It is called "Hey Siri" and allows you to use SIRI to control your devices hands-free. You can set it up so that you can use SIRI without having to physically touch your device. This allows you to give commands from across the room or from outside of the house. It will be used in conjunction with your Apple Watch if you have one because it can be used with it too.

Siri can group certain apps together for easier access.

"This is an app that allows you to group certain apps together for easier access. For instance, if you have several maps apps, you can create a shortcut for "Maps" and "Traffic" because they are both busy driving apps. This will combine them so that the two apps will appear next to each other on your home screen. This feature can be included in groups of a few shortcuts so that they can be easily accessed without having to individually tap them to launch them.

Siri has a unique feature called "Hey Siri." You must enable this feature before using it on your iPhone or iPad or the Apple Watch. In order to use "Hey Siri," you must have a device that has the option for it. This is usually an iPhone 6S or newer with iOS 11 installed and an Apple Watch, but it can also be the iPad Pro (2017) or the Mac Book Pro 2017. When using this feature, your iPhone will hear your commands without having to physically touch the device. In fact, you can talk to Siri from across the room if you place your connected devices in another room within your house.

Siri can lose its voice sometimes. If Siri stops working or you get a robotic response to your request, there are a few things that can help. You can check to make sure Siri is set to work on your device. If not, then you will need to access the "Hey Siri" feature from the iPhone settings and switch it off and then back on. Additionally, if Siri's voice still sounds strange when you ask for it, you can change its language in order to get a better response. For instance, if you are getting a British accent instead of an American one, you can use the "Voice" drop-down menu under General to choose between an American or British accent.

Siri Compatible Devices

There are many Apple devices that can use SIRI. The list is long and it includes the following:

iPhone XS, iPhone XS Max, iPhone XR, iPhone X (these devices will all be updated to iOS 12 when released in October 2018), iPhone 8 Plus, iPhone 8, iPhone 7 Plus, iPhone 7, the iPad Air 2 as well as the iPad 2017 and 2016. This means that you'll be able to use SIRI on your phone where ever you go as well as your tablet when at home or work. You'll also be able to access Siri from your Macbook Pro laptop.

Siri can be very useful for busy people. It's great for when you're in a rush and don't want to resort to using your phone's touch screens. You can make calls, find texts, schedule events, and more. Remember that there are some things that Siri will not be able to do. For instance, it won't tell you if it's going to rain or what's on TV tonight. But it will be able to tell you the traffic conditions on your way home from work or about your boss' meeting schedule for the day.

Siri can also be used in CarPlay compatible vehicles. This allows you to use your car's built-in voice system to give commands to Siri. An example of this is "Hey Siri, play the Beatles on Apple Music."

Siri can be used in the Apple Watch itself. This allows you to use your watch to control your phone's SIRI by asking it questions and giving commands from anywhere in the world. If you have an Apple Watch, you can also use Siri to send messages and make calls through your watch. You can also send outgoing messages that will be read out loud by Siri as well as receive incoming text messages through your watch.

After Apple releases its new iPhone models (iPhone 9, iPhone XS, etc.) in the fall of 2018, SIRI will be updated to use the new devices. Apple has promised that Siri will get better at responding to your queries and will also stop making mistakes. This is part of Apple's plan for improving SIRI over time to make it a more powerful tool for users.

Siri is based on Nuance's Dragon speech recognition system, which was originally developed by Bill Joy. It was used as the basis for Siri by Apple Inc., which bought the company in 2010 and released it as free software under certain conditions with iOS 7 (2011). The user interface and behavior of Siri were loosely based on Bing.

Modify Siri's Language

Siri offers a variety of languages for users to choose from. In order to access these different languages, you must first open the Settings app on your iPhone or iPad then tap on the General tab. Next, tap on Siri and select your language preference. By selecting your language preference, you are only changing the user interface of Siri. The actual voice recognition still comes from the original language setting that was created when you first set up Siri.

Modify Siri's Voice

Siri offers a variety of voices for users to choose from. In order to access these different voices, you must first open the Settings app on your iPhone or iPad then tap on the General tab. Next, tap on Siri and select your voice preference. By selecting your voice preference, you are only changing the user interface of Siri. The actual voice recognition still comes from the original language setting that was created when you first set up Siri.

You can also adjust some of Siri's settings in the Settings app on your iPhone or iPad.

Tapping on General then Siri will show you what Siri is currently set up for (language, voice, etc.).

Activate Siri's Announce Message feature.

To activate Siri's Announce Message feature, open the Settings app on your iPhone or iPad then tap on General then Siri. You will see a message that says "Siri Announcement Enabled." Tap to enable this feature if you wish to have ALL notifications announced by Siri—including phone calls even if you are driving.

Change Siri's Activation Method

To change the activation method for SIRI, open the Settings app on your iPhone or iPad then tap on General then Siri. You will see a message that says "Siri Activation Method." Tap to change this from the default setting of "Wake Word." Other activation methods include "Built-In Keyword" and "Manual."

Change Siri's Voice

To change SIRI's voice, open the Settings app on your iPhone or iPad then tap on General then Siri. Tap Voice Gender to change from Female to Male if you wish. Tap Voice to select a different voice from a list.

The default voice is "Voice 1 – English (United States)."

There are different voices for different languages (Chinese [Mandarin] has both male and female voices for example).

Change Siri's Language as well as Accent and Locale

To change SIRI's language, open the Settings app on your iPhone or iPad then tap on General then Siri. You will see a message that says "Siri Voice Language." Tap to change this from the default setting of "Default – English (United States)." Some languages are free while others require in-app purchases.

Change Siri's Formats (Only iOS 10+)

To change the formats that SIRI uses when it talks to you, open the Settings app on your iPhone or iPad then tap on General then Siri. You will see a message that says "Paraphrases." Tap to change these from the default settings of "My Choice" and substitute words with placeholder text like (name) or (birthday).

Change Siri's Voice Feedback (Only iOS 10+)

To change how SIRI speaks to you verbally when it understands you or is processing your request, open the Settings app on your iPhone or iPad then tap on General then Siri. You will see a message that says "Voice Feedback." Tap to change these from the default settings of "Always" and change what SIRI says when it understands, can't help right now, has something else to do, and can give feedback later.

Siri for iPhone is compatible with the following iOS versions:

iOS 5 or higher.

Siri for Mac is compatible with the following macOS versions:

Siri has been criticized by many users. A common criticism became that it started all of its answers by saying "That's an interesting question" instead of starting conversations more intelligently. This led to Siri being nicknamed "The useless assistant." It also had a reputation for not understanding jargon, slang, and slang abbreviations, which could often lead to confusion in answering questions, as well as not providing details on places and events in urban environments, often leading to users asking their iPhones if they had directions when they did not.

Announce Messages with Siri' can be managed.

Siri does not play music on many older versions of iOS.

Siri was criticized for its interaction in China, where users are often required to provide their ID number or fingerprint identification to use Siri. The user then often has to answer several questions just to have their request processed by Siri.

These issues were resolved with updates that made Siri localizable, removing concerns such as this in other languages and countries, but the OS still requires “Intelligent Assistance” access on these systems.

In December 2014, Apple updated Siri with a new interface inspired by iOS 7. At the same time, it was also revealed that "Siri's core technologies were being enhanced so that Siri can understand more natural language and context." Prior to this update, the voice recognition used for Siri was based on deep learning and text analysis rather than statistical analysis of previous inputted requests. This made the speech recognition much more accurate and helped it become a lot better in understanding what users were trying to say. Users' requests could be sent straight to apps installed on their iPhones or iPads. The update added third-party support, allowing users to send requests to apps that have integrated with Siri.

In May 2016, another update was made for Siri. The update allowed for third-party messaging apps to be used directly through Siri. Additionally, a new feature called "SiriKit" was revealed. This allows for a variety of third-party apps to integrate with Siri in a more streamlined way in order to allow users more control over those apps from their voice and other commands from the home screen and lock screen on iOS 10 or macOS Sierra.

In September 2016, Apple announced an upcoming update to Siri including increased use of machine learning in order to provide better responses and suggestions as well as much more advanced search capabilities.

CHAPTER 12:

iCloud

iCloud is an internet service that connects all of your Apple devices and keeps them in sync. Here's all you need to know.

iCloud is Apple's most current (and longest-running) online service for its subscribers, including email, calendars, contacts, content sync, and online storage similar to Google Drive or Dropbox. The service is available on iOS and macOS and Windows and the web, and there are many storage levels to select from.

Apple's iCloud service keeps your digital life in sync across all of your devices. Everything from calendar events to your most recent vacation picture album can be stored and synced in iCloud, so you can view it from any Apple device.

iCloud also supports synchronization for various third-party applications for iPhone, iPad, and Mac, ensuring that documents, lists, podcasts, and other items are always up to date on the devices in question.

What Is iCloud+?

Apple unveiled iCloud+, a new enhancement to their iCloud service, at WWDC 2021. Unlike the company's previous plus-branded service offerings, iCloud+ is more of a brand of iCloud itself than an extra tier or premium add-on.

Hide My Email and Private Relay, as well as enhanced HomeKit Secure Video storage, are among the new features in iCloud+. If you have a premium iCloud subscription, you already have iCloud+ and can use these additional features for free.

Is iCloud Free to Use?

You can use iCloud for free, but even if you have the best iPhone or iPad, you might not get the best experience.

An iCloud account is free to create and use, but it comes with a modest (and honestly, insufficient) amount of storage: 5GB. Your images, movies, iCloud Drive content, and iCloud backups will rapidly use this data. You won't be able to add new stuff to your iCloud storage once it's full unless you delete older files or upgrade to a premium iCloud storage plan.

What Are the Costs of iCloud Storage Plans?

You can choose from a number of different storage plans in iCloud, and the addition of iCloud+ hasn't changed the costs or storage tiers. Here is the breakdown.

- 5 GB of free storage
- $0.99 per month for 50GB
- $2.99 per month for 200GB
- $9.99 per month for 2TB

If you wish to update, increase, or downsize the amount of storage you have for your iCloud account, you can do it at any time using one of your Apple devices.

Is iCloud Included with Apple One?

Yes, Apple One, Apple's service package, contains iCloud and Apple Music, Apple News+, Apple TV+, Apple Fitness+, and Apple Arcade.

The amount of iCloud storage provided depends on the Apple One package you choose: Individual, Family, or Premier, which each gives 50GB, 200GB, or 2TB of iCloud storage. If you use many of Apple's other services, it's probably a good idea to combine them with Apple One to save money.

What Is an iCloud Backup?

In essence, an iCloud backup is an off-device backup of your device's data. You'd be able to rapidly have a replacement phone set up with all of the same content and information if yours was lost, stolen, or destroyed. Regularly backing up your iPhone or iPad (or any device, for that matter) is a smart practice if the worst happens.

If your iPhone or iPad is connected to Wi-Fi and charged, it will back up to iCloud automatically every day by default. Most people's phones will go through this process overnight, which is when they are most likely to satisfy these requirements for a prolonged length of time. You can also back up your iOS devices manually at any time using the Settings app.

When restoring or setting up a new iOS device, you may opt to restore from a recent iCloud backup, which will restore everything on the device from that date and time.

What Exactly Does an iCloud Backup Entail?

Apple's official list of what's included in an iCloud backup is as follows:

- App data
- Device settings
- Apple Watch backups
- iMessage, text (SMS), and MMS messages

- Home screen and app organization
- Videos and photos on your iPhone, iPod touch, and iPad

Your purchase history is stored in music, movies, TV shows, applications, and books purchased via Apple services.

Ringtones

Password for Visual Voicemail (requires the SIM card that was in use during backup).

Only the data on your iOS device is included in your backups. Contacts, Notes, Calendars, your iCloud Photo Library, Reminders, and iCloud Drive content are all synced with iCloud and don't need to be backed up. Even if you set a device up "as fresh," i.e., not from a backup, you will still see this sort of synchronizing data when you sign in with your iCloud account.

Where Can I Use iCloud?

Since its launch in 2011, every current Apple product has had iCloud support. You may manage your iCloud account in the Settings app on iPad, iPhone, and Apple TV, and as System Preferences on macOS. You may also use iCloud.com to manage everything and recover certainly lost stuff.

Can I Access iCloud on the Web?

Yes, you can. You can use any contemporary browser to access Mail, Calendars, Contacts, Reminders, Notes, and Photos on iCloud.com. You can also look at what's on your iCloud Drive, as well as the web-based versions of the iWork suite, Numbers, Keynote, and Pages.

Each of these web apps has a lot of the same features as its iOS and macOS versions, such as picture management in Photos, event creation, even document creation, managing events in

Calendars, and collaborating with others on projects in the iWork apps.

Furthermore, iCloud.com works well with Find My to help you track down missing devices or see where your friends and family are.

Can I Use iCloud on Windows?

Yes, you certainly can. iCloud is fully supported on Windows, and you can set up iCloud syncing by downloading the official iCloud for Windows program to your PC.

iCloud for Windows can sync files to iCloud Drive, transfer photographs to your iCloud Photo Library, sync your contacts, mail, and calendar with Outlook, and even sync your bookmarks with your preferred browser.

Can I Use iCloud on Android?

There is currently no official way to sync contacts and calendars, save photographs in the iCloud Photo Library, or keep content in iCloud Drive using your iCloud account with Android devices.

However, you can use your iCloud email address with numerous email apps on Android, so you can at least get your email messages if you do must utilize an Android device. Android web browsers can now access iCloud.com.

Can I Share iCloud Storage with My Family?

Yes, you definitely can. In addition to sharing an Apple Music subscription and App Store purchases, your family members will be able to access your iCloud storage if you set up Family Sharing. When you choose this option, Apple emphasizes that all photographs and documents will remain private, and family members will not see what others have saved on the iCloud.

How Do I Sign Up for an iCloud Account?

You can create a new iCloud account on any iPad, iPhone, or Mac. You will be offered to utilize an existing Apple ID for iCloud or create a new Apple ID for iCloud when you set up a new device.

If you don't already have one, you can quickly create a new Apple ID on your iPad or iPhone or set up a new Apple ID on your Mac and use it with iCloud.

Information Apple Keeps or Syncs With iCloud?

iCloud keeps almost all of Apple's native iOS applications and their Mac counterparts in sync. These include:

- Mail
- Calendar
- Notes
- Contacts
- Messages
- News
- Reminders
- Health
- Home
- Wallet
- Maps
- Shortcuts

- Books
- Find My

You can also use the iCloud Photo Library to sync your photographs. Safari bookmarks and open tabs can be synchronized across all of your iCloud devices with Safari installed, allowing you to continue surfing the web on any of them. iCloud can be used to safely store and sync your log-in information for multiple websites using iCloud Keychain, too.

If you need basic online storage, iCloud Drive will save and sync everything you put into it, making documents, photographs, music files, and much more available across your iOS and macOS devices. As previously stated, iCloud can sync data from third-party applications to keep your complete digital life up to date across all of your devices, and third-party apps can also utilize iCloud Drive for storage.

How You Can Find Your Lost iPhone

Find My app could not be active without being initially activated through the Settings App. Therefore, for you to make the feature of Find My App to be functional do the following steps:

- Find friends and family members
- Share your location with others
- Set Up Find My

How You Can Activate Find My on Your iPhone

Hint: "Find My" is automatically turned on when you Sign Into your new iPhone with your Apple ID, but there is a need for you to find out if Find My iPhone, Enable Offline Finding and Send Last Location are activated.

Home Screen: Hit on Settings.

Hit on your Name/Sign in on your iPhone beside the profile picture.

Apple ID: Hit on Find My.

Hit on Find My iPhone to select "On" if the activation feature is off.

Turn on the activators of the following if they are off:

- **Find My iPhone:** It will always request your Password to locate, erase, or lock your iPhone.
- **Enable Offline Finding:** Your iPhone will be located even when it is not connected to a cellular or Wi-Fi network.
- **Send Last Location:** iCloud will automatically send the location of your iPhone to Apple when the battery is drastically low.

Hit on My Location to select This Device.

Hit on Share My Location's Activator to put on the switch.

What to Do after You Have Lost Your iPhone

There is a possibility of misplacing your iPhone in a location that you could not recollect because you had visited more than three places before you could remember that your iPhone is missing or it was stolen by a thief.

Also, you might have kept the iPhone in a compartment that is best known to you only but after a while, you could not remember specifically where exactly you had hidden the iPhone.

However, Apple has made a reliable way of locating your iPhone with the use of Find My iPhone and iCloud Map Detection or Google Map to specifically describe and identify the location of your iPhone wherever it had been kept.

First Finding Solution

On Mac or PC:

Use any available browsers such as Chrome, Mozilla Firefox, Internet Explorer, Safari, Opera, Lynx, or Konqueror.

In the Web Add text field type icloud.com.

iCloud Homepage: Sign in with your Apple ID which includes your email and password.

Click on the Find iPhone icon among the icons on the screen.

iCloud Find My iPhone: Location Map will display on the screen.

At the top bar center of the screen click on All Devices.

On the drop-down, select your iPhone (if you are using more than one Apple device that is using the same Apple ID. You will see all the Apple devices on the drop-down).

The Map will zoom out to indicate the iPhone location with a black circle and your iPhone' Name label.

In the right corner, you will see 3 options you can use to help your findings: Play Sound, Lost Mode, and Erase iPhone.

Play Sound: If you are pretty sure that the iPhone is around (home, workshop, or office) then you can click Play Sound immediately you will start hearing vibration with sound. The sound will continue until you hit on Find My iPhone Alert OK.

Lost Mode: This option is perfect when you discovered that you can no longer recover/find your iPhone again, then you can click on Lost Mode.

Phone Number Dialog Box: Type your Phone Number (that can be called by the finder) and click on Next at the top right corner of the dialog box.

Click on the Text box and type the message that will be shown on your iPhone screen. For example, "Please I have lost this iPhone. Kindly call me. Thank you" and click on Done at the top right angle of the dialog box. Automatically the iPhone will be locked. It will only be unlocked if you enter your passcode or through your Face ID.

Erase iPhone: This option is accurate when you realized that the iPhone could not be located, then, click Erase iPhone to completely remove all your vital and confidential data, applications, and documents from the iPhone quickly.

Second Finding Solution

This can be used when you are having Google Map App on your iPhone but if do not have the app you can use the iCloud method above.

On Mac or PC:

Use any of the available browsers such as Chrome, Mozilla Firefox, Internet Explorer, Safari, Opera, Lynx, or Konqueror.

In the Web Add text field type www.google.com/maps

Google Maps: Click on the Menu Icon at the top left side of the Search Google Maps Text Field.

Menu: Select your timeline (timeline will show various locations you have been with your iPhone on Google maps with an indication of red color).

On the left side of the screen click on Today at the front of the Timeline.

Immediately a line will displace your different movement today on Google maps to know where your iPhone could be found.

You can zoom in on the map to make the location to be closer and clearer for you to see the location very well.

You can increase the displacement line (appear in blue) on the left side of the screen to see more different places with their specific time in hours, minutes, and seconds that you moved from a particular place to another place till the final place where the iPhone could be found.

Note: You won't see the exact spot where the iPhone could be found on the Google map but you could only know the environment where you can see the iPhone.

How You Can Use another iPhone to Track Your Lost iPhone

The method of tracking your lost iPhone on the computer is virtually the same as the method and processes of locating your lost iPhone on another iPhone. If you are having two iPhones in your family or your friend is having one, you can easily use the available iPhone to quickly discover where your iPhone is kept or the location at which it could be found with either hope of recovery or not.

Home Screen: Hit on the Find My app icon.

- **Find My:** Hit on the Devices icon at the bottom center of your iPhone.
- **Devices:** Hit on your lost iPhone's Name (e.g., Ephong's iPhone) that is your first name will be used to qualify the lost iPhone.

- **iPhone's Name:** Swipe up from the top edge center of the page to select one of the available options below:
- **Play Sound:** Having 100% assurance of finding it.
- **Directions:** If it is discovered on the map and you want it to be tracked down.
- **Notifications:** You will be notified when the iPhone is found.
- **Mark as Lost:** If the recovery chances of your iPhone is 70–80% but the location on the map displayed is extremely far. Then you can hit on Active.
- **Erase This Device:** If the chance of recovering your iPhone is less than 50% which is very narrow, then you can choose the option by tapping it.
- **Hint:** If you eventually selected Erase This Device because you have initially lost hope of finding it, as a result, you would not be able to track your iPhone again. But, if by slim opportunity you found the iPhone, you will only be able to restore all the erased data and documents on your iPhone through iCloud backup.

How You Can Share Your Location with Others

You will need to select the people that you wanted to share your location with. This option will enable you to track down your lost iPhone on your friend's iPhone.

Home Screen: Hit on the Find My app icon.

Find My: Hit on the People icon at the bottom left of the screen. **People:** Hit on Start Sharing Location Start Sharing Location.

Type the Name of the Person you want to share your location with into the Text Field and hit on Send at the top right angle of the page. An Optional Dialog Box will show up to select "time for the sharing of your location with the person." Select any of the flowing options: Share for One Hour, Share Until End of Day, or Share Indefinitely.

Notification: On a Dialog box you will see "You Shared Your Location with the Person's Contact. Hit on OK.

How to Use Your iPhone IMEI Code to Block Your Lost iPhone

Your iPhone IMEI Code is very important to you when your iPhone has been confirmed lost and no hope of recovery.

It is very essential to know your iPhone IMEI code which serves as the main personal Apple device identification.

You could use your IMEI code to confirm if your iPhone is originally manufactured by Apple Company or produced by a fake manufacturer. It can also be used to unlock your iPhone.

Therefore, you have to take the following wise step to confirm your iPhone IMEI code:

Home Screen: Hit on the Phone app.

Phone Page: Dial Asterisk "", Hash "#", Zero "0", Six "6", Asterisk "", Hash "#" (i.e., #06#) on the keypad.

Tap on the Call button.

Immediately, the IMEI code will appear on your iPhone screen.

Keep the IMEI Code in a confidential planner.

Call your Phone Operator, report the lost iPhone, and dictate your iPhone IMEI code for the lost iPhone to be blocked.

CHAPTER 13:

Advice and Recommendations

Save Seconds in Your Searches

Concerning learning the footy ratings or proving an area, getting where you need to be on the web is targeted on speed and precision. Something is missing if you're forced to knock out type-heavy websites. To save lots of period by keeping down the complete quit icon while keying in and out an address to speak about a short-cut band of URL suffixes. From your classics (.com, .co.uk) to the less used (.edu, .ie), you will find quick-strike shortcuts for all people. You might have submitted away the Compass application alongside the Shares and found out Friends applications inside a folder entitled "Crap I cannot delete." You should draw it out again since it offers an integral second function that will aid together with your DIY responsibilities. Instead, swiping the left in the Compass application introduces an advantageous level—an electric bubble measure that can check if that shelf is level.

Lock Your Camera's Center Point

Everyone knows that tapping the display while taking a photo will set the camera's point of concentration, right? Good. Annoyingly, though, every time you move the camera after choosing a center point, it disappears. Instead of simply tapping the screen, press for another or two until an "AF Locked" container arises. You may twist, change, and swing finished around without dropping focus.

Create Custom Vibrations

Ever wished you could show who's calling by just how your phone feels buzzing against your lower leg. You can, in Connections, select your person of preference and strike Edit. Here you will notice a Vibration option. Selecting this offers you many choices, just like the Create New Vibration tool. Making your bespoke hype will be as simple as tapping the display towards the defeat of the decision.

Snap a Picture without Pressing Your Phone

An oldie, but a goodie iPhone hack, is utilizing your volume control buttons to capture simple, thus maintaining your meaty paw blocking the screen as you make an effort to strike the touchscreen settings. However, if you want a lot more, remove your photo-capturing shutter handles, hitting the quantity button on a couple of compatible, connected earphones could have the same impact.

Keep Your Data Allowance by Restricting App Access

You're only another of simply halfway through the month, plus your 2GB data allowance was already beginning to appear slightly extended. You don't have to scale back on your own on-the-go Netflix looking. Instead, choose which applications get demoted towards the Wi-Fi-only B-list. Check out Configurations > Mobile Data, where you can create the very best decisions on the application simultaneously.

Improve Your Battery Pack Life

Limelight, Apple's linked quick-access for key data and services, is fantastic for offering access immediately to the most recent breaking information, sports ratings, and social upgrade. But quite definitely, stuff happening in the backdrop can consume your battery pack life whole. Unless you switch off Limelight features for some applications to obtain even more life per charge, that's. Just proceed with Configurations > General > Limelight Search and limit what's attracting data behind your consent.

Improve Your Transmission by Knowing Where You Can Search

You don't have to walk out initial floor windows looking for where your iPhone's connection is most appropriate. Type *3001#12345#* into the iPhone's dialer and strike call release on the concealed Field Setting tool.

This sub-surface menu becomes your pub chart-based signal indication into a far more simple numerical-based sign signified. Got a rating of -50?

You then will exist enjoying H-video streams on the highway. However, if it drops around -120, it will battle to send a wording. Just observe the elevated indicator figures.

Discover Out How Much Time You Have Been Getting Excited About a Chat

We've all been there, endlessly rechecking our mobile phones for a wording reply, thinking how long it's been since we sent our message of love.

Finally, there's a straightforward way to understand it.

Swipe in from your right-hand area of the display when within a messaging thread, showing exact delivery occasions for every message delivered and received. True, it is not as morale-beating as WhatsApp's blue ticks, but it will give you complications over why it's overtaking 42 minutes for your other 50% to reply. Do affairs take that long?

Share All Your Family Members' Tree with Siri

Does discussing your parents with by given names make you look awkward? Then train Siri to understand whom you're communicating with. Ask Siri to call your father, and the digital PA should ask who your dad is. Once a contact continues to be designated towards the parent moniker, you will be supported without problems every time you require pops to continue.

Increase a Slow iPhone

Computing devices tend to decelerate after a while as components degrade, storage fill with old documents and overlooked/unused apps, and new software is increasingly designed for newer and faster processors.

You can defer the inevitable by following some simple guidelines, including:

- Every occasionally, you should power off your device completely; this clears out the storage.
- It's also advisable to enter the habit of deleting applications and files you don't use (photos certainly are a universal problem for space for storage) and archive the latter in the cloud and local backup.
- It's also really worth going through the configurations and checking which application refreshes in the backdrop, thereby burning treasured control power.

Upgrade iOS on Your Device

Understand that updating iOS has historically been a combined blessing when it comes to accelerating your iPhone, but with iOS, they have changed.

iOS is particularly devoted to performance. For example, Apple stated it might get old devices faster, and it appears to do so within our tests.

More tips can be found on how best to increase a slow iPhone with this book and other books by Engolee Publishing House.

Outset Dark Mode

If your iOS (during writing, it's available as a general public beta; it'll launch standard in Sept 2019), you can rapidly transform the dark settings system. Thus giving all the pre-installed applications and any third-party applications, which have built-in compatibility a dark or dark-grey background that's more calming to learn at night.

To start Dark Setting, start the Settings application carefully, and tap Screen & Brightness. Close to the surface of the following display, you will notice Light and Dark options hand in hand—tap the one you intend to use. You can also specify Dark Setting to read, sometimes, from dusk until dawn automatically.

Skip Calls With Remind Myself Later

Alternatively, you can have iOS remind you to return the call later. As with the auto-replies, how you do this depends on your version of iOS—you probably press the Remind Me button above the glide; in earlier versions, you had to swipe up before you could choose to Remind Me Later.

Instead, you can get reminded in 1 hour, “When I Leave” or (where applicable) “When I Go Back Home.” Ensure that your address details are currently in Contacts, meaning your iPhone understands where home is usually. The timings depend on your Navigation movements.

Produce an iPhone Safe for Kids

Kids love iPhones, but you can find actions you may take to make sure children cannot access unsuitable content for the devices.

Check out Settings > General > Limitations, and you'll limit using specified apps, in-app buys, and place quite a while for appropriate articles; all this can be protected in How to create parental controls with an iPhone.

It's also sensible to browse the odds of Family Sharing, an attribute that lets you talk about applications content between your family's devices and not have to purchase them more than once.

The arrival of iOS provides further parental controls utilizing Screen Time, which lets you set 'allowances' for the usage of certain applications or types of apps, warnings when time is running out, and lastly, a block. (They can ask for additional time, but you will get the best decision).

End iPhone Addiction

Talking about Display Time, it's a new characteristic in iOS that may help you be less dependent upon your iPhone. To understand the quantity of time and effort you are wasting on your own iPhone, head to Settings > Display Time.

Here you will notice information about how much time you have on each app, how often you view your cellular phone, and which applications you spent the most time with. Touch on your device in the most effective section to see the Display screen Period breakdown.

You can test the breakdown for today or go back seven days. You can set a Screen Time account password to use if you prefer a few more moments.

Promptly Add Symbols

You have likely been with your iPhone's keyboard for a long period without realizing that it's easier than you thought to add icons to your communications.

Instead of tapping once in the 123 buttons, once on your own selected symbol, and once more in the ABC button to come back to the original keyboard layout, you can certainly do the entire lot in one gesture.

Tap the 123 buttons, slip your finger to find the sign you intend to place, then release. Once it's been added, your keypad will automatically revert; one tap instead of three: that's some serious time cost benefits right there.

Oh, despite the fact that we're talking symbols: maintain your finger on any notice or mark for another or two, and you'll see how many other (usually related) icons the fact that button can offer instead. For instance, the buck key offers pound, euro, and yen icons. If you often type words with accents, this is also an instant and easy way to see an accented option.

You'll find so many additional symbols hidden within your keyboard that you may don't have discovered. Experiment!

One-Handed Keyboard

This feature is feasible if you're on iOS or later versions.

Check out Settings > General > Keyboards and tap One-Handed Keypad. Select Left or Right.

The iOS QuickType system-wide keypad is clever at guessing what you are endeavoring to create and, in many situations, will autocorrect your typed being somewhat much more accurate. However, when you start customizing it to learn your private favorite shortcuts and abbreviations and the complete phrases you desire it to expand those abbreviations into.

You might decide that "omg" ought to be changed into "Oh my God," for instance, "omw" should become "On my way," etc.

You can make a personalized shortcut:

- Head to Settings > General, scroll down and touch Keyboard.
- Select Text Alternative; you will notice what text replacements you have set up.

Enable Nighttime Shift

Nighttime Change dims the screen's white shades to make it easier for your eye in low-light circumstances. You can regularly make Night Shift happen simultaneously each day, or manually allow it until tomorrow.

You can also adjust the color temperature so that it is just about warm.

Check out Settings > Screen and Lighting > Night Change.

Suppose you are long-press for the speech bubble within your Messages. In that case, you'll now come across the decision to 'Speak'—the decision is especially useful when you have a protracted text or opt to begin traveling and want to listen to the written text while in a hands-free setting.

Watch Whenever You Get a Message

In the Messages app, you can swipe forth left to expose the time-stamps of each message.

Usually, you can see the date using the period when the first message was sent; however, to reveal every individual message after that, you will need to feel the timestamps by swiping promptly. This is good for knowing what period the final call was received or to get out if your friend was lying about arriving on time.

Call From Within Messages

If you're chatting via Messages and decide it might be beneficial to talk instead, you can simply tap the icon of the person you're chatting with to call.

Apple may not have changed the visuals all that much with iOS 15, but there is still enough to get used. Whether you've just purchased a new iPhone 13 or have recently upgraded your current iPhone. Most of the highlighted features are available on all iPhone models running iOS 15, although a few are exclusive to the iPhone 12 and iPhone 13 series and won't work on previous devices. So, ensure your iPhone is running iOS 15—or get your iPhone 13/13 Pro—and try out some of our favorite, less-obvious features. You can either watch the video tutorial or follow the written instructions below. Whichever option is most convenient for you.

1 - Stop Applications from Tracking You

Many apps wish to track various patterns and use data from you while using their applications. Most of the time, it is so they can serve you more relevant adverts or learn how people use their app so they can improve it. However, there are compelling reasons to disable all tracking, particularly if the corporation has a history of violating your privacy rights.

Go to Settings > Privacy and then hit "Tracking." There is a button at the top that reads, "Allow applications to request to track." Toggle this off if you don't want any applications to request tracking information. Alternatively, you can manually turn on or off tracking for each app in the list below.

2 - See Which Applications Access What Data

Here's a quick hint on a similar topic: "Record App Activity" can be found at the bottom of the Privacy > Settings page. Turn it on and leave it for a week or so. Return to that menu option after a few days to get a breakdown of how various apps have used permissions on your phone and see if any of the apps have had access to things like your location or other sensitive information much too frequently. You can also save a report as a blank binary file that you can save or share.

3 - Bring the Old Safari Back

The upgraded Safari in iOS 15 has a tab switcher at the bottom and a search bar. It's extremely handy for one-handed usage and allows you to switch between tabs by swiping quickly. You can, however, revert to the original style, which includes the URL bar at the top.

It may be done in two ways. Alternatively, go to Settings > Safari and scroll down until the two tab icons appear. Select "Single tab" now. The bar will be at the top when you reopen Safari.

The alternative option is to launch Safari, go to any site, and then hit the "AA" button in the search box. You will now see a pop-up menu appear. Select "display top address bar" from the drop-down menu.

4 - Bundle Non-Important Notifications

Instead of receiving tons of random pings throughout the day, iOS 15 has a "Scheduled Summary" feature that allows you to gather all of the less important notifications together and give them all at once.

Go to Settings, then Notifications, and select "Scheduled Summary" to activate it. To finish the setup, switch it on and follow the on-screen instructions. Now choose the applications you want to include in your summary and hit the "Add Apps" button at the bottom of the screen.

On the following screen, choose the times you wish the first and second summaries to appear. Adding more frequent summaries during the day is also possible by using the "+" button. If you like, you could have one come at breakfast, lunch, and then again in the evening. When you have made your selections, go to the bottom and choose "Turn on Notification Summary." You will only be notified via direct messages or applications that aren't included in your selection until you select to have the summary appear.

5 - Announce Notifications on Airpods

Siri can voice notifications through your AirPods Pro when you have them in your ears and connected to your iPhone. To turn this on, head to Settings > Siri and Search, then tap "Announce Notifications" at the top of the page and turn it on. Turn on the "Headphones" setting.

6 - Disable HDR Video

By default, the new iPhone 13 Pro models record video in Dolby Vision or HDR, which takes up a lot of storage and isn't suitable with a lot of other devices. To turn it off, go to Settings > Camera and select "Record Video." Toggle the "HDR Video" option off.

7 - Reachability

When iPhones had buttons, you could use a gentle double press on the home button to bring items from the top of the screen down to make them easier to reach. You can do that on the new phones without buttons, but you must first activate it.

Tap "Touch" in Settings > Accessibility. Toggle the Reachability option on or off. When you swipe down from the top of the screen, the items from the top are now dropped lower down, making them simpler to access.

8 - Double Tap the Back to Screenshot

To take a screenshot, just tap twice on the back of your phone. Instead of fiddling with the button combination. This will be if the appropriate accessibility option is enabled.

Simply go to Settings > Accessibility > Touch and pick "back tap" from the drop-down menu. Select the double-tap option and select Screenshot from the drop-down menu. When you double-tap the rear of your phone, it will now take a screenshot. You can, of course, choose additional features such as notifications or the control center, which might be difficult to access.

9 - Use Live Text

Live Text is an excellent new feature that lets you read and show text in real-time using your camera. You could, for example, use it to transfer a sentence from an article or document directly into a message.

Open a texting app on your phone. Double-press the text field, then touch the little symbol that looks like text in a box. It will open a camera view and begin scanning text, displaying it in the field as you show it to the camera.

To be more specific, touch the little capture symbol in the corner to quickly grab the texts you want. You can now highlight the precise block of text you want to include by dragging your finger over the text you want.

You can also achieve this by just opening the camera app and selecting Photo mode. Now press the live text icon in the corner, then point to the text you want to focus on or tap to manually focus on the text. It will scan the text, and you'll be able to highlight text inside the popup picture to share with someone. Copy and paste to your desired location.

10 - Live Text in a Screenshot

Live text can also be used differently. To open up the editor view, take a screenshot and tap it. Toggle between the markup/pen icon at the top and the live text icon in the bottom corner. You can now drag to highlight the text you wish to copy and paste into any app.

11 - Focus Mode

The Focus mode could easily be its in-depth feature and video. However, you can create dedicated situations where only specified applications and individuals can disturb you at specific times or places by going to Settings > Focus.

You can create two scenarios: one for personal use and another for professional use. This could take time to complete the procedure, but it mainly adds dedicated people who can contact you and select which applications to authorize. It's a really simple process. Once that's done, you can set a timer for it to turn on. It doesn't matter whether it's based on time and date or place. You can even modify the appearance of the home screens that show during each focus period.

12 - Limit the Screen Frame Rate

The iPhone 13 Pro and 13 Pro Max's displays are set to a maximum refresh rate of 120Hz by default. If you want to restrict anything for whatever reason, you can, but you won't find it under the Display options. Go to Settings > Accessibility. Toggle on the "Limit Frame Rate" option and will not exceed 60 fps.

13 - Create Your Personal Photographic Style

Photographic Styles is another camera function available on the latest iPhone models. Unlike a basic filter, this one modifies factors like warmth, contrast, and brightness to create a look that will be applied to all of your images as soon as you take them. Open Settings > Camera, then scroll to "Photographic Styles." Swipe through the different options until you find one you like. Once you've found one, touch the blue tap at the bottom.

14 - Change the Music on Your Photos Memories

Apple automatically adds a background track to the picture and video memories it makes for you in the "For You" tab of the new Photos app. However, you can change this. Open Photos and touch "For You," then choose one of the Memories at the top of the screen.

Touch it again after it's open, then tap the little music icon in the corner. You can now swipe across the screen to adjust the filter and music. Alternatively, you can hit the musical note icon in the bottom right corner to reveal a bigger list of selections, including a search button that, when clicked, will search your whole library for a certain song.

15 - Trackpad Keyboard

Finally, but certainly not least. This is an old favorite, but it's a must-try. To move your cursor when typing, you can transform your keyboard into a trackpad by just long-pressing the space bar and then flicking your thumb around. Now place the cursor on the screen where you want it to go.

CHAPTER 14:

Solutions to Common Problems

Most challenges encountered with the iPhone can easily be resolved by restarting your device. However, we will look at every possible challenge you may have with the iPhone and the solutions.

Complete iPhone Reset Guide

Most minor issues that occur with the iPhone can be resolved by restarting the device or doing a soft reset. If the soft reset fails to solve the problem, then you can carry out other resets like the hard reset and master reset.

How to Restart/Soft Reset iPhone

This is by far the most prevalent solution to many problems you may encounter on the iPhone. It helps to remove minor glitches that affect apps or iOS as well as gives your device a new start. This option doesn't delete any data from your phone, so you have your contents intact once the phone comes up. You have two ways to restart your device.

Method 1:

- Hold both the side and Volume Down (or Volume Up) keys at the same time until the slider comes up on the screen.
- Move the slider to the right for the phone to shut down.

- Press the Side button until the Apple logo shows on the screen.
- Your iPhone will reboot.

Method 2:

- Go to Settings then General. Click on Shut Down.
- This will automatically shut down the device.
- Wait for some seconds, then Hold the Side button to start the phone.

How to Hard Reset/Force Restart an iPhone

There are some cases when you need to force restart your phone. These are mostly when the screen is frozen and can't be turned off, or the screen is unresponsive. Just like the soft reset, this will not wipe the data on your device. It is important to confirm that the battery isn't the cause of the issue before you begin to fore-restart.

Follow the steps below to force-restart:

- Press the Volume Up and quickly release.
- Press the Volume Down and quickly release.
- Hold down the Side button until the screen goes blank and then release the button and allow the phone to come on.

How to Factory Reset Your iPhone (Master Reset)

A factory reset will erase every data stored on your iPhone and return the device to its original form from the stores. Every single data from settings to personal data saved on the phone will be deleted. You should create a backup before you go through this process. You can either backup to iCloud or iTunes. Once you have successfully backed up your data, please follow the steps below to wipe your phone.

- From the Home screen, click on Settings.
- Click on General.
- Select Reset.
- Chose the option to Erase All Content and Settings.
- When asked, enter your passcode to proceed.
- Click Erase iPhone to approve the action.

Depending on the volume of data on your phone, it may take some time for the factory reset to be completed.

Once the reset is done, you may choose to set up with the iOS Setup Assistant/Wizard where you can choose to restore data from a previous iOS or proceed to set the device as a fresh one.

How to Use iTunes to Restore the iPhone to Factory Defaults

Another alternative to reset your phone is by using iTunes. To do this, you need a computer, either Mac or Windows, that has the most current version of the iOS as well as has installed the iTunes software. A factory reset is advisable as a better solution to major issues that come up from software that wasn't solved by the soft or force restart. Although you will lose data, however, you get more problems fixed, including software glitches and bugs. Follow the guide below once all is set:

- Use the Lightning cable or USB to connect your device to the computer.

- Open the iTunes app on the computer and allow it to recognize your device.
- Look for and click on your device from the available devices shown on iTunes.
- If needed, chose to back up your phone data to iTunes or iCloud on the computer.
- Once done, tap the Restore button to reset your iPhone.
- A prompt will pop up on the screen, click Restore to approve your action.
- Allow iTunes to download and install the new software for your device.

I Can't Activate My Mobile Phone

There are four possible reasons for this. Causes and solutions are listed below:

If the activation was not done correctly, repeat the activation steps as instructed in this book.

There may be a temporary problem with your network connection. Wait for some time and try again or you can move to any location and then try again.

The problem may be with the SIM card. If so, contact your carrier's customer service for assistance.

Owners of the iPhone 13, the iPhone 13 mini, the iPhone 13 Pro, and the iPhone 13 Pro Max are reporting a range of glitches and performance difficulties.

We'll walk you through some of the most frequent iPhone 13, iPhone 13 mini, iPhone 13 Pro, and iPhone 13 Pro Max issues in this article.

Our list covers possible solutions for Wi-Fi troubles, Bluetooth issues, and a variety of other frequent iPhone 13 issues.

Before you begin fiddling with your settings, ensure that your iPhone 13 is running the newest version of iOS 15. New software is always capable of addressing bugs and performance difficulties.

How to Fix Battery Life Problems

The majority of iPhone 13 customers report receiving outstanding battery life from their devices. However, some users report that the battery drains more quickly than it should.

If the issues are severe, you may have a hardware problem. If that is the case, you should contact Apple customer support and/or visit an Apple Store with your iPhone 13.

Instead, you can:

Restart Your Mobile Device

- Update turn off 5G on your iPhone
- Examine your applications
- Reset all configurations
- Utilize the low-power mode
- Put an end to background app refresh

How to Resolve Bluetooth Issues

If your iPhone 13 is unable to connect to one or more Bluetooth devices, there are a few possible solutions.

To begin, you'll want to disable the Bluetooth connection that isn't functioning correctly.

This is how you do it:

1. Go over to the Settings app.
2. Bluetooth should be tapped.
3. Utilize the I in the circle to choose the connection.
4. "Forget this Device" should be selected.
5. Reconnect the Bluetooth device if necessary.

If it does not resolve the issue, you might attempt to reset your network settings:

1. Go over to the Settings app.
2. Select General.
3. Reset the device.
4. Reset Network Settings by tapping on it.

This procedure should take no more than a few seconds to finish. Additionally, it will lead your iPhone to forget recognized Wi-Fi networks, so keep your password(s) available.

Additionally, you can attempt to restore your iPhone 13's settings to their factory defaults. This should be a last resort. This is how you do it:

1. Go over to Settings.
2. Select General.
3. Reset the device.
4. Select Reset Transfer or Reset iPhone from the menu.
5. Reset the device.
6. Reset All Settings by tapping Reset All Settings.

7. If you have activated a passcode, enter it.

If none of these options resolve the issue, you may need to contact Apple's customer care. If the product you're attempting to connect to is not manufactured by Apple, you'll want to contact the manufacturer.

How to Resolve Wi-Fi Issues

If you're experiencing poor Wi-Fi speeds or lost connections, here are a few possible solutions that have previously worked for us.

Before you begin fiddling with the settings on your iPhone 13, you should analyze the Wi-Fi connection that is creating the issues. If you're connected to your home Wi-Fi network, reboot your router.

Whether you're certain there are no difficulties with the router itself, visit Down Detector to see if others in your region using the same internet provider are experiencing similar troubles.

If you're unable to reach the router to which your iPhone 13 is connected, or if you're certain the issue is not related to your ISP/router, open the Settings app on your iPhone 13.

Here, you'll want to forget about the problematic Wi-Fi network. This is how you do it:

1. Tap Wi-Fi in the Settings menu.
2. By touching the I in the circle, you may select your link.
3. At the top of the screen, tap Forget this Network (note that this will lead your iPhone to forget its Wi-Fi password, so keep it handy).

How to Resolve Cellular Data Issues

If your iPhone 13 suddenly shows a "No Service" message and you are unable to connect to your cellular network, follow these procedures.

To begin, verify that your neighborhood is not experiencing a power outage. Check for reports on social media and/or connect with your service provider on social media. Additionally, you may visit Down Detector to see if others in your region are experiencing similar problems.

If the issue is not connected to a network outage, you should restart your iPhone 13 to see if it resolves the problem. If it does not work, try activating Airplane Mode for 30 seconds and then deactivating it.

If this still does not work, you may want to try totally turning off Cellular Data. To accomplish this, follow these steps:

1. Go over to the Settings application.
2. Activate Cellular.
3. Toggle off Cellular Data.
4. Allow it to cool for a minute before turning it back on.

How to Resolve Sound Issues

The iPhone 13's speakers should be capable of producing loud, clear sounds. However, if your sound begins to crackle or sound muffled, try the following before contacting Apple customer care.

To begin, reboot your iPhone 13. Additionally, you may want to verify that your SIM card is properly positioned in the tray. The iPhone 13's SIM card slot is positioned on the left side of the device.

If the sound is still absent or distorted from the phone, check for debris covering the speaker grille or the Lightning port on the bottom of the device.

Restart your iPhone if you notice an abrupt decline in call quality. Additionally, you'll want to examine the device's receiver for obstructions like dirt or your screen protector (if you have one). Additionally, you may try removing your case (if you have one) to see if it helps.

If your phone's microphone suddenly ceases to function or begins to cut out sporadically, consider restarting your iPhone 13.

If the microphone remains broken, you can attempt to restore it from a backup. If restoring fails, you'll want to contact Apple, since you may have a hardware issue.

How to Resolve Face ID Issues

If you're experiencing trouble using Face ID on your iPhone 13, here are a few things to try.

To begin, verify that your iPhone 13 is running the most recent version of iOS 15. If you're on the most recent version of iOS 15 and experiencing problems, navigate to your Face ID settings.

1. Go over to Settings.
2. Proceed to the Face ID and Passcode section. Take note that you will be required to enter your passcode (if applicable) in order to obtain access.

Once inside, verify that Face ID is configured on your phone and that all of the functions for which you're attempting to utilize Face ID are now enabled.

If you're having trouble unlocking your phone with your face, make sure you're gazing at the screen actively.

If your look is continually changing, you may need to add an alternate appearance to Face ID. How to configure an alternate appearance:

1. Go over to Settings.
2. Select Face ID and passcode.
3. Set an Alternate Appearance by tapping on it.

Additionally, you'll want to ensure that no material (dirt, dust, etc.) is obstructing your iPhone 13's FaceTime camera.

If your phone does not recognize your face during the Face ID setup process, ensure that you are scanning it in a well-lit area. Additionally, you may need to pull the iPhone closer to your face or level it.

Conclusion

Although the Apple iPhone phone is intended to be intuitive, it can still be very difficult for an average user to understand how to use the features effectively. As you can see, this guidebook was created with this in mind as it covered everything you need to know about the iPhone to get started.

With the conclusion in mind that you want to handle your iPhone seamlessly, it's important to take note that the iPhone's touch screen will get dirty, particularly on the edges and corners. It's important to clean your iPhone with a microfiber cloth regularly. You'll find these in the cleaning aisle of your local grocery store.

There are many tips and tricks to help you maximize your iPhone experience. The more time you spend learning about your new phone, the more you'll be able to get out of it. So use the information in this guidebook to make the most of your device and enjoy the extra benefits that come with having a top-of-the-line smartphone!

In conclusion, the iPhone is an incredibly powerful device that has a lot of different uses. From taking photos or making videos to organizing your schedule and staying connected with friends, family, and colleagues. The ease of use and design of the iPhone makes it a well-rounded device for people of all ages. If you take care of it, making sure it stays far from water or excessive dust and that the screen is kept clean and free of scratches, it should last you a long time.

The iPhone has a lot to offer no matter what your age. It has an amazing camera, an incredible processor, and a long-lasting battery making it perfect for everyday use. With so many features that are specifically tailored to seniors, there's no question that this smartphone will help you in day-to-day life.

The iPhone is available at all cell phone carriers and can be bought with a contract or on its own plan. The plan for the iPhone will differ in price depending on the type of plan you get.

You've now completed this guide on the iPhone, congratulations! You're now ready to take your new device out into the world. Take some time to familiarize yourself with it and make sure you take care of it so it lasts a long time. Remember, even though it's relatively well built, not everything is meant to last forever!

I am confident that you will be able to enjoy to the fullest all the fantastic features of the iPhone. The iPhone is not only a phone for making and receiving calls. With the right knowledge of using the iPhone, you can achieve more incredible things with this device.

Printed in Great Britain
by Amazon

14705547R00112